Research on the Seismic Design of Steel Frames

Research on the Seismic Design of Steel Frames

Guest Editors

Linfeng Lu
Shaofeng Nie

MDPI

Basel • Beijing • Wuhan • Barcelona • Belgrade • Novi Sad • Cluj • Manchester

Guest Editors
Linfeng Lu
School of Civil Engineering
Chang'an University
Xi'an
China

Shaofeng Nie
School of Civil Engineering
Chang'an University
Xi'an
China

Editorial Office
MDPI AG
Grosspeteranlage 5
4052 Basel, Switzerland

This is a reprint of the Special Issue, published open access by the journal *Buildings* (ISSN 2075-5309), freely accessible at: https://www.mdpi.com/journal/buildings/special_issues/4959693Y2D.

For citation purposes, cite each article independently as indicated on the article page online and as indicated below:

Lastname, A.A.; Lastname, B.B. Article Title. *Journal Name* **Year**, *Volume Number*, Page Range.

ISBN 978-3-7258-4839-3 (Hbk)
ISBN 978-3-7258-4840-9 (PDF)
https://doi.org/10.3390/books978-3-7258-4840-9

Contents

About the Editors

Linfeng Lu

Linfeng Lu, Ph.D., is a Professor at Chang'an University and the head of the research team "Design Theory and Application of Steel (Composite) Structures and New Structural Systems." He serves as a standing member of the Testing, Appraisal, Strengthening and Renovation Subcommittee, and the Aluminum Structure Subcommittee of the China Construction Metal Structure Association. His teaching and research focus on the fundamental theory and practical application of steel structures. He is the chief editor of the first English textbook on "Fundamentals of Steel Structures" published in China. He has led more than ten national and provincial-level research projects, including those funded by the National Natural Science Foundation of China. He has also directed major engineering design projects such as the reconstruction of the Yongning Gate Watchtower in Xi'an, the Dingding Gate Museum in Luoyang, and the Xiongjiazhu Tomb Museum in Jingzhou. He has published over 100 academic papers and received prestigious honors, including a Second Prize of the National Science and Technology Progress Award and an F5000 Leading Paper Award. His recent research interests include innovative structural systems for steel structures, corrosion and durability of steel structures, and strengthening and retrofitting techniques.

Shaofeng Nie

Shaofeng Nie, Ph.D., is a Professor at Chang'an University and the Director of the Cold-Formed Steel Research Center. He is recognized as a Young Science and Technology Star of Shaanxi Province and serves as a member of the Prefabricated Building Expert Committee of the Department of Housing and Urban—Rural Development of Shaanxi Province. His primary research interests include the fundamental theory of steel structures, light-gauge steel structures, steel–concrete composite structures, and structural wind engineering. He has presided over more than 20 research projects, including the National Natural Science Foundation of China, the Shaanxi Key Research and Development Program, and other provincial-level scientific and technological initiatives. He has also participated in over 30 additional research projects. He serves as a youth editorial board member of several academic journals, including Journal of Architecture and Civil Engineering, Journal of Chang'an University (Natural Science Edition), and Journal of Transportation Engineering. He has published more than 80 academic papers, including over 30 indexed by SCI and EI. He has received the First-Class Award for Scientific and Technological Progress from the China Steel Construction Society.

Preface

This Reprint commemorates the 40th anniversary of the 1994 Northridge earthquake, a pivotal event that profoundly reshaped the seismic design philosophy for steel structures worldwide. The collected papers reflect ongoing efforts to improve the safety, resilience, and performance of steel and composite structures under seismic actions. Topics include innovative joint configurations, such as end-plate and panel zone connections, seismic behavior of cold-formed and concrete-filled steel elements, and performance optimization using energy dissipation devices. Both experimental and numerical approaches are featured, offering a comprehensive view of current advances in earthquake-resistant structural engineering. This collection aims to serve as a valuable reference for researchers, engineers, and code developers committed to advancing seismic safety. We express our sincere appreciation to all contributing authors and reviewers for their dedication to this important topic.

Linfeng Lu and Shaofeng Nie
Guest Editors

buildings

MDPI

Article

The Influence of Box-Strengthened Panel Zone on Steel Frame Seismic Performance

Linfeng Lu *, Jichen Zhang, Gaowei Zhang, Hao Peng, Bingyou Liu and Hanlin Hao

School of Civil Engineering, Chang'an University, 75 Chang'an Middle Rd, Xi'an 710062, China; 2023228008@chd.edu.cn (J.Z.); 2023228047@chd.edu.cn (G.Z.); 2023228056@chd.edu.cn (H.P.); 2023228019@chd.edu.cn (B.L.); 2021128031@chd.edu.cn (H.H.)
* Correspondence: lulinfeng@chd.edu.cn

Abstract: During the Northridge earthquake, extensive brittle failures on the weld zones of the beam bottom flanges in the rigidity connection of steel special moment frames (SMFs) were detected. One of the primary reasons is the high-tensile strain demand created at the beam bottom flange zones due to positive bending. The weak panel zone of the I-section column exhibits more shear deformation, which promotes and accelerates the brittle fracture of the beam bottom flange weld zones. A box-strengthened panel zone can minimize the shear deformation of the panel zone of the I-section column, which may also reduce the inter-story displacement of steel SMFs and enhance their seismic behavior. In order to investigate this fact, in this research we carried out a model test of a steel frame with a box-strengthened panel zone to examine SMFs' seismic performance and inter-story displacement, as well as testing the contribution of panel zone shear deformation to inter-story drift. Numerical methods were then used to investigate the influence of the axial compression ratio and beam-to-column linear stiffness ratio on the effect of shear deformation on the box-strengthened panel zone. Design recommendations are given based on the research results.

Keywords: steel frame; seismic performance; panel zone; shear deformation; story drift

1. Introduction

The 1994 Northridge earthquake in California significantly impacted seismic steel research, design, and construction practices, particularly concerning steel special moment frames (SMFs) and their performance during earthquakes, which has raised concerns regarding the reliability of the pre-Northridge connection method and prompted engineers to rethink the design of connections and construction details of connections in steel moment frames. However, all of the various types of connections are based on two concepts: (i) strengthening the connection and (ii) weakening the beam ends. The overall goal in developing new connections is to provide a highly ductile response, reliable performance, and economy.

Before the Northridge earthquake, steel SMFs were widely considered one of the most effective systems for earthquake resistance. These frames were designed to produce ductile responses during earthquakes. This ductile behavior was expected to occur through flexural hinges at the beam-to-column connections and shear yielding in the column panel zone, resulting in a ductile plastic mechanism in the frame. However, many multi-story steel structures with SMFs during the earthquake had brittle fractures in the beam-to-column moment connections [1–5]. These incidents generated severe concerns about the design and construction of SMFs, resulting in a design revolution in steel frame joints and connections. Steel frame connections have since been classified as pre- and post-Northridge connections.

Researchers conducted tests and other investigations to determine the causes of the failure of the pre-Northridge connections and constructed several post-Northridge connections to prevent brittle earthquake failure [5]. Koetaka et al. [6] proposed a novel weak-axis column bending connection with hysteretic dampers. Their test demonstrated that the proposed connection achieved stable hysteretic performance in an extensive deformation range because yielding was limited only in the dampers. Cabrero and Bayo [7,8] performed an experimental and a theoretical study to investigate the behavior of extended end-plate connections in both major and minor column axes subjected to a three-dimensional loading. The tested minor-axis joints consisted of two partial end plates outside welded to the column flange. Similarly, Kim et al. [9] proposed three new weak-axis connection types with welded split and end plates. They reported that these types were easy to construct and ensured the flexural behavior of weak-axis moment connections. This weak axis connection was tested and reported again by Lee et al. [10]. Lee et al. [11,12] investigated the seismic performance of six types of weak-axis column-tree connections through cyclic testing of six full-scale specimens. The effects of beam splice length on the seismic performance of weak-axis column-tree connections were experimentally investigated. Additional work has been conducted by Shim et al. [13]. They proposed a weak-axis system, which mainly used bolts, as shown in Figure 1. In their research, eight interior joint specimens were tested to verify the structural behavior of the proposed connection, and their test results demonstrated that it behaved better than the standard weak-axis connection and had excellent constructability. Kozlowski [14] presented a comprehensive analytical model to predict the moment resistance, initial stiffness, and rotation capacity of semi-rigid weak axis composite connections, and comparative analysis with test results showed a good correlation.

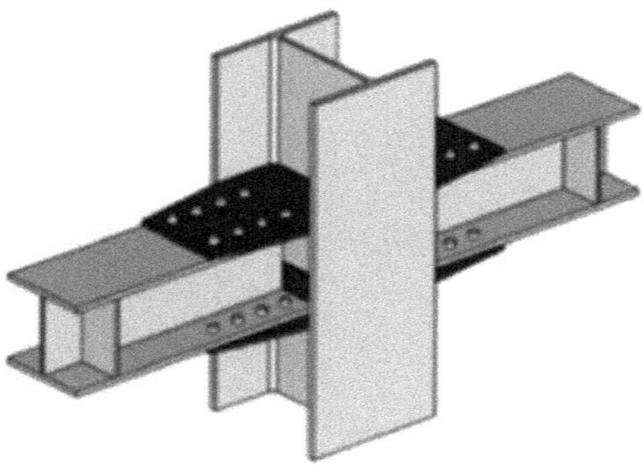

Figure 1. Connection proposed by Shim.

Kim et al. [15] illustrated a composite beam-to-column joint to explain that one of the reasons for failure is the concrete slab, which causes the neutral axis to move toward the top flange when the beam is subjected to positive beam bending (concrete slab under compression), causing the strain of the lower flange to be much greater than that of the top flange and leading to premature connection brittle failure. In addition to the issues with the concrete slab, the shear deformation of the panel zone can also harm the strain of the lower flange of the beam, leading to excessive demand for tensile strain and causing the weld to fracture between the beam's lower flange and the column. For example, Adlparvar et al. [16] studied the impact of panel zone shear strength on the seismic behavior of enhanced slotted-web beam connections, and their results indicated that high participation of the panel zone effect increased the likelihood of beam web weld fracture. Miri et al. [17] also found that in frames with weak panel zones, the story drifts are more than the permitted rate according to design standards, and the story drifts could be reduced by reinforcing the panel zone. Therefore, for an SMF with I-section columns, in a typical weak panel zone steel frame, it was necessary to provide substantial inelastic deformation capacity via flexural yielding in the beams and restricted elastic deformation in panel zones.

The steel SMF's seismic design is often based on two concepts: high ductility–low bearing capacity or low ductility–high bearing capacity. However, according to the rules of the specification GB50011-2010 [18], the seismic design of SMFs should follow high ductility–high elastic bearing capacity for strong columns and weak beams and strong joints and weak components. When SMFs fail during an earthquake, plastic hinges emerge on the beam ends, but no plastic failure occurs in the columns or panel zones. To realize this design concept, a novel type of joint must combine a strong panel zone with high ductility and bearing capacity. In 2018, Lu et al. [19] proposed a box-strengthened panel zone joint for SMFs with I-section columns, which were typical weak panel zone steel frames, as shown in Figure 2, and gave the detailed construction procedure and methods of the developed connection. It has good rotational ductility, its plastic rotation capacity is more than 0.03 rad, and its bearing capacity is slightly higher than that of the box-section column [19]. It also has strong panel zone characteristics and can effectively achieve seismic functions such as plastic hinge displacement at the beam end [19]. Thus, this study analyzes the influence of this new panel zone on the seismic performance and story drifts of steel SMFs by tests and numerical analyses to promote the engineering use of this new panel zone connection.

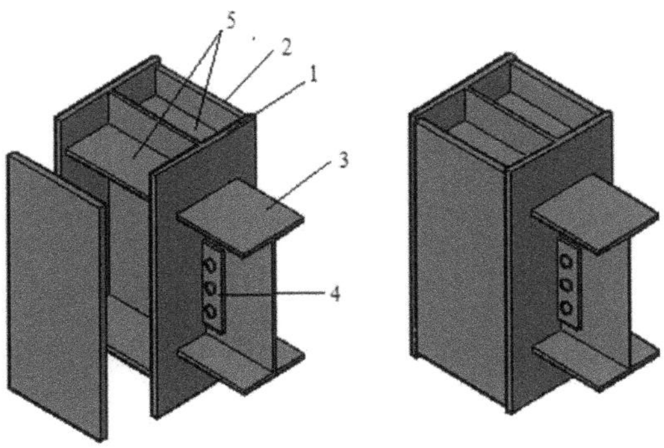

Figure 2. The box-strengthened panel zone joint. 1. I-section column; 2. Skin plate (pair); 3. H-shaped beam; 4. Shear plate; 5. Diaphragm (two pairs).

2. Design Specification for the Box-Strengthened Panel Zone Joint

In order to avoid the recurrence of the Northridge earthquake damage, it is essential to consider the combination effect of concrete slabs and propose technical measures to reduce panel zone shear deformation in the design of the post-Northridge connection. Due to the complexity of the failure mechanism of composite joints, even when some design codes adopt steel–concrete composite beams, the design of steel frame connections, considering the influence of concrete slabs, has not been included in existing specifications. On the contrary, some specifications accept strengthening SMFs' panel zones [17–19]. The two specifications, GB50011-2010 [18] and ANSI/AISC 341-22 [20], both require the seismic design of the panel zone, including the required shear strength, the panel-zone thickness, and the requirements of panel-zone doubler plates, which are used if the thickness of the column web does not meet that shown in Equation (1) [18,20].

$$t \geq (d_z + w_z)/90, \tag{1}$$

where $d_z = d - 2t_f$ of the deeper beam at the connection, d is the total depth of the beam in. (mm), and t_f is the thickness of the beam flange, in. (mm); t = thickness of column web or individual doubler plate, in. (mm); w_z = width of panel zone between column flanges, in. (mm).

When plug welds are used to join the doubler plate to the column web, the total panel-zone thickness is permitted to satisfy Equation (1). Additionally, the individual thicknesses of the column web and doubler plate shall satisfy Equation (1), where d_z and w_z are modified to be the distance between plug welds. When plug welds are required, a minimum of four plug welds shall be provided and spaced in accordance with Equation (1).

However, except in Equation (1), the limit state of shear yielding needs to be checked according to Equation (2) [18].

$$\Psi(M_{pb1} + M_{pb2})/V_P \leq (4/3)f_{yv}, \quad (2)$$

where M_{pb1}, M_{pb2} = the fully plastic bending bearing capacity of beams on both sides of the panel zone, respectively; Ψ = reduction factor; f_{yv} = shear yield strength; V_P = volume of panel zone; for the I-section column, $V_P = d_z w_z t$, and for box-section column, $V_P = 1.8 d_z w_z t$.

The volume of the panel zone of the box-section column should be double that of the I-section column, which was reduced by 10% when implemented in GB50011-2010 [18]. The author proposed the construction of the box-strengthened panel zone of the I-section column depicted in Figure 2 regarding the shape of the box-section column panel zone. The volume of the panel zone of the I-section column shown in Figure 2 is greater than that of the box-section column when the thickness of the skin plate is the same as that of the box-section column wall plate because it has an extra column web in its panel zone. Existing research has demonstrated that the stiffness of a box-strengthened panel zone joint is greater than that of the box-section panel zone joint [19].

3. Experiment Investigation on the Seismic Behavior of a Steel SMF with a Box-Strengthened Panel Zone

Of course, according to the design specifications [18,20], adding the doubler plates will also increase the volume and stiffness of the I-section column panel zone and reduce the story drifts [17]. However, the design specification GB50011-2010 requires that the calculation of an SMF's story drift when it uses I-section columns must consider the shear deformation of the panel zone, while for an SMF that uses box-section columns, it is not necessary. In frames characterized by vulnerable panel zone areas, in 2009, Miri et al. [17] observed that when story drifts exceed the permissible limits defined by design standards, reinforcing the panel zone with doubler plates can effectively mitigate the development of excessive story drifts, but the deficiency of their research is that they did not quantitatively analyze the contribution of shear deformation in the panel zone strengthened by doubler plates to the story drifts. In order to investigate the influences of shear deformation on the story drift of SMFs, an SMF specimen experiment was conducted.

3.1. Experiment Specimen

We designed a Q235 steel SMF specimen, as shown in Figure 3. The column was HW300 × 300 × 10 × 15, and the beam was HN350 × 175 × 7 × 11. The thickness of the skin plate was 16 mm, the thickness of the diaphragm was 12 mm, and the thickness of the shear plate was 8 mm. Beam top and bottom flanges and column flanges or skin plates were fully penetrated bevel butt-welded with an E4311-type welding electrode; the beam web and the connection plate welded to the column flange or skin plate were connected by six 10.9-grade M20 high-strength bolts. The mechanical properties of steel plates, welds, and H-shapes are listed in Table 1.

Table 1. Member sections and analysis results.

Item	f_y/MPa	f_u/MPa	E/MPa	δ/%
Column web, 10 mm	278.7	460.6	206 162	28.4
Column flange, 15 mm	253.1	472.1	208 780	27.1
Beam web, 7 mm	332.2	468.0	202 324	27.8
Beam flange, 11 mm	295.5	443.2	204 381	29.1
Skin plate, 16 mm	275.0	443.3	207 000	29.7
Diaphragm, 12 mm	295.0	443.2	206 130	28.9
Shear plate, 8 mm	305.5	440.4	206 600	29.3
Welds	398.3	496.7	208 000	18.8

f_y is measured yield strength; f_u is measured tensile strength; E is measured modulus of elasticity; δ is elongation ratio.

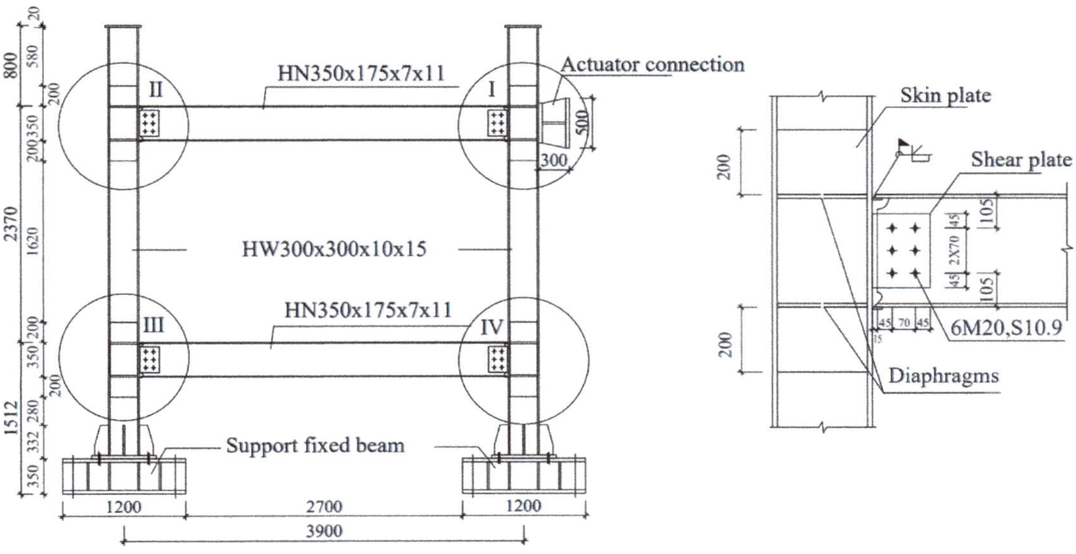

Figure 3. The box-strengthened panel zone SMF and joint details.

3.2. Loading Setup and System

Figure 4 depicts the test loading configuration. The specimen's columns were connected to the support fixed beam via a stiff exposed column base secured to the ground with a compression beam. Two anchor bolts secured the pressure beam to the base channel. Four horizontal lateral supports were installed on both sides of the beam in the first and second stories of the specimen to avoid lateral-torsional buckling and out-of-plane bending of the beam.

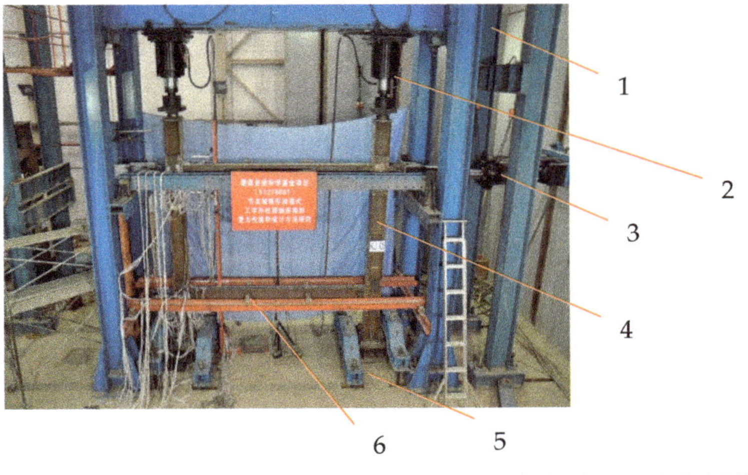

Figure 4. Test setup. 1. Loading portal steel frame; 2. Electro-hydraulic servo jack; 3. MTS actuator; 4. Steel frame specimen; 5. Compression beam; 6. Lateral support. Note on the red signboard: Nature Science Foundation of China (51278061), research on the mechanical performance and design methods of the weak axis connection of the box-strengthened panel zone of the I-section column.

An MTS actuator with a maximum thrust of 1460 kN, an ultimate tensile force of 960 kN, and a stroke of 250 mm was used for horizontal loading. An electro-hydraulic servo vertical loading gantry system with a maximum loading capacity of 3000 kN was used for vertical loading. The data were gathered using the IMP static data gathering equipment. A vertical force was applied in four stages at the top of the column based on the axial compression ratio of 0.45. After each load level was used and stabilized for three minutes, we checked whether the load value was constant and eliminated the installation gaps of the loading device and the specimen. The first stage used 286 kN, the second used 573 kN, the third used 859 kN, and the fourth used 1146 kN. The vertical force of 1146 kN remained constant throughout the test. Through finite element numerical simulation before the test, the calculated yield load displacement was confirmed as 20 mm. Therefore, after vertical loading of 1146 kN, the MTS actuator applied a low cycle repeating load corresponding to the displacement step. The loading steps are shown in Figure 5. There was a displacement of 4 mm in each direction per loading step with a loading rate of 0.5 mm/s and three cycles at each step. After the horizontal load dropped to 85% of the peak load, we continued loading for one more cycle and stopped the test.

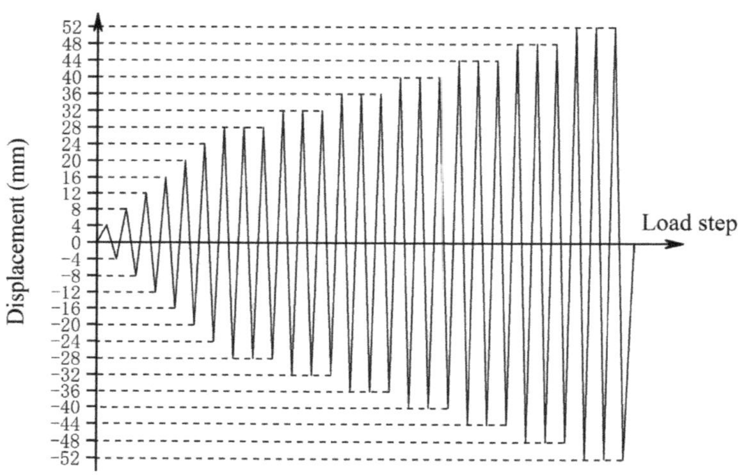

Figure 5. The loading steps.

3.3. *Displacement Meter Arrangement*

Figure 6 depicts the experiment's layout of displacement meters and strain gauges. D1 and D2 displacement meters detected the horizontal displacement of the beam on the second and first floors, respectively. The horizontal displacement of the top and bottom of the specimen column was measured by displacement meters D3 and D4. The horizontal movement of the support fixed beam was measured using a displacement meter D5. In order to describe the test phenomenon easily, this research divided the nodes of the frame specimen into four areas, marked with Roman numerals I, II, III, and IV, respectively, as shown in Figure 3. The rotational deformation of II and III beam-column joints was measured using displacement meters D6, D7, D8, and D9. The rotational deformation of II and III columns was measured using displacement meters D10, D11, D12, and D13, respectively. At the appropriate places, displacement meters D14, D15, and D16 were used to measure the vertical deflection of the second-story beam; displacement meters D17, D18, and D19 were used to measure the vertical deflection of the first-story beam.

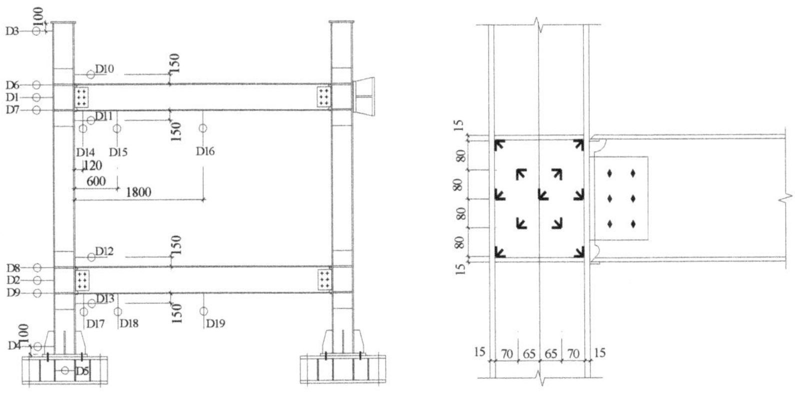

(a) (b)

Figure 6. Displacement meter and strain gauge arrangement: (**a**) displacement meters; (**b**) strain gauges.

The strain gauges were arranged in the panel zones (skin plates) of areas II and III, and a typical arrangement is shown in Figure 6b. The strain gauge measured the shear strain in the panel zone and monitored its stress state (elasticity, elastoplasticity, or plasticity) when the specimen was damaged.

3.4. Test Phenomena and Failure Characteristics

When a horizontal displacement of 28 mm was applied, the strain approached yield. The top flange of the beam end I buckled in the first cycle when loaded to 48 mm (Figure 7a), and the column base flanges buckled somewhat in the second cycle. The lower flange of the beam end II buckled when loaded to 52 mm (Figure 7b). When loaded to 56 mm, the web of the beam end I buckled in the first cycle, the upper column flange of the beam end IV buckled in the second cycle, and the middle part of the beam cracked at about 30 mm from the column flange at the lower flange of the beam end II (Figure 7c) and the upper part of the column flange III buckled in the third cycle. The lower flange at the beam end II shattered entirely after the second cycle when loaded to 60 mm. The top flange of the beam end I split when loaded to 64 mm, and the beam web bulged in an entire waveform (Figure 7d). The top flange of the beam end I entirely shattered when loaded to 68 mm, causing the specimen to fail. Figure 8a depicts the overall failure morphology, and Figure 8b shows the panel zone's stresses at the peak loading state in area II. The horizontal displacement for this peak loading condition was 56 mm.

The specimen yielded first at the upper flange of beam end I because the I joint was closer to the loading point than the others when all of the specimen's frame beams met the bending rigidity. Nonetheless, the specimen fractured sequentially at the lower flange of beam end II and the upper flange of end I, both of which were steel plate fractures. During the loading process of the test, the ball joints of the left and right column top jacks appeared to be locked, causing the force on part II of the specimen to be significantly greater than that on part I, which led to the accumulation of tensile damage and the cracking of the steel plate. The upper flange steel plate at the beam end I fractured when the tensile displacement reached 64 mm, and the failure position was roughly symmetrical with the center of the beam's lower flange failure position. This indicated that the energy released by the fracture of the lower flange of beam end II was diagonally transmitted to the upper flange of beam end I.

Figure 7. The local failures of the specimen: (a) the beam upper flange buckling in area I; (b) the beam lower flange buckling in area II; (c) the beam lower flange fracture in area II; (d) the beam upper flange fracture and web buckling in area I.

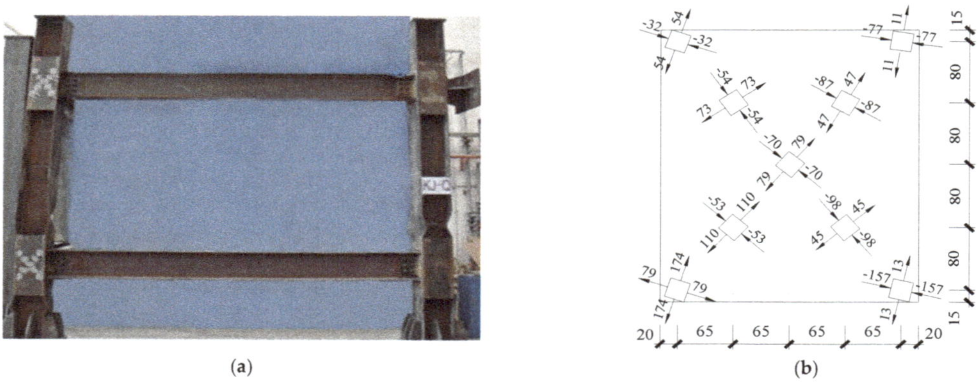

Figure 8. The failure mode and panel zone stress of the specimen: (a) the overall failure when the horizontal loading displacement was 68 mm; (b) the stress of the panel zone in area II when the horizontal loading displacement was 56 mm (unit: MPa).

Figure 8b demonstrates this: (1) The stress in the panel zone was often low, indicating that the panel zone remained elastic even at peak load and did not contribute to energy dissipation. This supported the earlier conclusion that a box-strengthened panel zone is strong. (2) The stress at the panel zone measuring locations was generally distributed along the diagonal direction, with one diagonal direction in tension and the other in compression. This demonstrated that shear deformation occurs in the panel zone, although it is elastic. (3) The principal stresses were quite high at the two corners of the panel zone closest to the beam flange, and the direction deviated from the diagonal direction to some amount. This demonstrated that the tensile and compressive stresses of the beam flange butt weld had a substantial impact on the stress distribution in the panel zone.

3.5. Seismic Behavior Analysis

The specimen's hysteretic and backbone curves are shown in Figure 9. In the pictures, P denotes the actuator's load (pushing forward and pulling back), and Δ represents the inter-story displacement of the specimen. Figure 9a demonstrates that the specimen is in the elastic phase early in the experimental loading, and the hysteretic curve is almost straight. As the loading displacement rises, the hysteretic curve fills out, forming a shuttle shape. This shows that, despite the box-reinforced panel zone strengthening the frame, the frame still had a high energy dissipation capacity. Local buckling occurred at the column base flange sites of the left and right columns during the loading process, resulting in a reduction in the lateral stiffness of the specimen during the mid-term (corresponding to loading displacement in the period of 20 mm to 40 mm) of the experimental loading. The hysteretic curve revealed a pinching trend later in the experiment due to the abrupt drop in specimen stiffness produced by the consecutive fracture of the second-story beam's top and lower flange plates. According to FEMA273 [21], the specimen's yield load, maximum load, and ultimate load, as shown in Figure 9b, and their associated inter-story displacements were estimated using the backbone curve of the specimen, as given in Table 2. Table 2 shows that the specimen's reverse bearing capacity was greater than its forward bearing capacity, but its reverse displacement ductility coefficient was less than its forward displacement ductility coefficient. This is consistent with the test's observation of early fracture in the opposite direction.

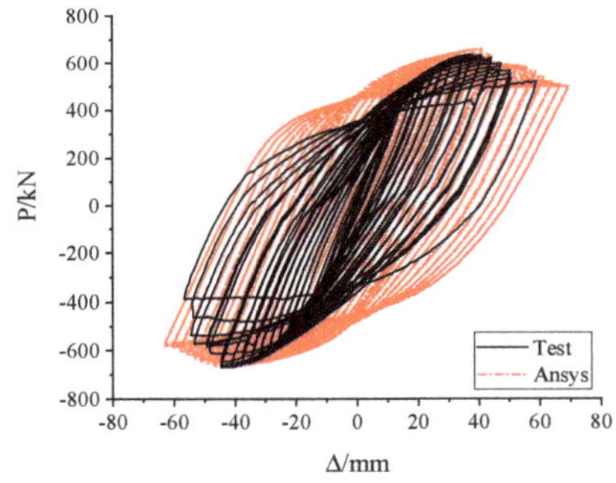

(a)

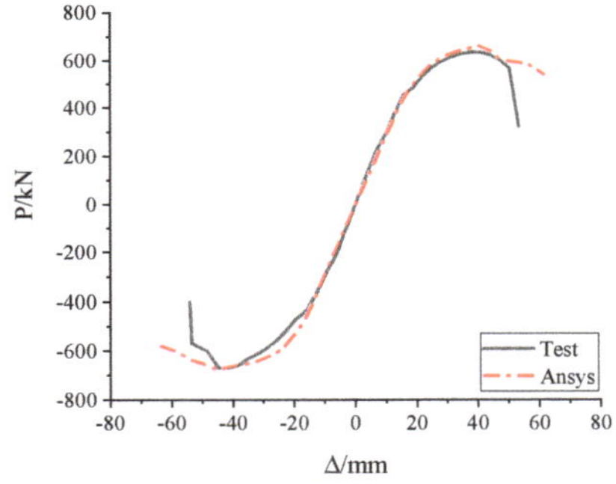

(b)

Figure 9. The curves of the specimen: (**a**) hysteretic curve; (**b**) backbone curve.

Table 2. Test results.

Loading Direction	P_y/kN	Δ_y/mm	P_{max}/kN	Δ_{max}/mm	P_u/kN	Δ_u/mm	Ductility Coefficient
+	481.2	18.9	638.1	37.9	542.4	55.6	2.94
−	472.7	19.9	669.7	44.7	569.2	53.4	2.68

P_y, P_{max}, and P_u represent the yield, maximum, and ultimate loads of single-bay frames, respectively. Δ_y, Δ_{max}, Δ_u is the inter-story displacement corresponding to P_y, P_{max}, P_u.

According to the relationship between the inter-story displacement angle and the inter-story displacement, θ = Δ/h, it was calculated that when the specimen reached the yield load, the corresponding inter-story displacement angle was 1/125 and was greater than 1/250; when the specimen reached the failure load, the corresponding inter-story displacement angle was 1/43 and was greater than 1/50. They all met the requirements of specification GB 50011-2010 [18].

3.6. Panel Zone Shear Deformation

Through the strain gauges arranged on the diagonal line of the panel zone, the strain value along the diagonal line of the panel zone of the specimen during the loading process was measured. The shear deformation angle of the panel zone during the test was calculated by Equation (3).

$$\gamma = \frac{\sqrt{a^2+b^2}}{ab}\overline{X}, \tag{3}$$

where a is the height of the panel zone, b is the width of the panel zone, and $\overline{X}$ is the average strain value in the diagonal direction.

In comparison, the theoretical shear deformation angle was calculated by Equation (4) following the specification GB50011-2010 [18].

$$\gamma_i = \frac{1}{n}\sum \frac{M_{j,i}}{GV_{\text{pe},ji}}, (j = 1, 2, \cdots n), \tag{4}$$

where $G = 7900\ \text{kN/cm}^2$ is the shear elasticity modulus of Q235 steel, V_{pe} is the effective volume of the panel zone calculated by Equation (2), and $M_{j,\,i}$ is an unbalanced moment of j-th panel zone in the i-th floor.

Table 3 shows the shear deformation angle of the panel zone within the elastic stress range of the specimen. Among these values, Δ_t is the horizontal displacement imposed during the test, Δ_1 represents the approximate frame inter-story displacement without considering the shear deformation of the panel zone, which was calculated according to the real-time value of the displacement meters, Δ_2 represents the frame inter-story displacement that was accurately measured considering the shear deformation of the panel zone, and Δ_3 is the frame inter-story displacement according to GB 50011-2010 [18]. The shear deformation angle of the panel zone was calculated according to Equations (3) and (4). In Table 3, γ represents the panel zone's yield shear deformation angles calculated by Equation (3), and γ_c represents the panel zone's yield shear deformation angles calculated by Equation (4).

Table 3. Comparison of frame inter-story displacement and panel zone shear deformation.

Δ_t/mm	Δ_1/mm	Δ_2/mm	Δ_3/mm	Δ_n/mm	$\gamma/10^{-3}$ rad	$\gamma_c/10^{-3}$ rad	γ/γ_c/%
4	3.1	3.119	4.2	3.3	0.008	0.4	2
8	6.3	6.336	8.4	6.6	0.015	0.9	1.7
12	9.5	9.557	12.4	10.0	0.024	1.2	2.0
16	12.6	12.676	16.4	13.3	0.032	1.6	2.0
20	15.8	15.897	20.4	16.6	0.041	2.0	2.1
24	18.7	18.828	23.6	19.8	0.054	2.1	2.6
28	22.1	22.256	27.5	23.2	0.066	2.2	3.0

Table 3 shows that the specimen's Δ_2 was similar to the Δ_1, which is the inter-story displacement without considering the shear deformation of the panel zone, and the increase in Δ_2 over Δ_1 was only 0.6% to 0.7%, showing that the contribution of the shear deformation of the panel zone was almost equal to zero. That means the shear deformation of the box-strengthened panel zone hardly enhanced the frame's inter-story displacement. Moreover, the actual shear deformation of the panel zone γ was only 3% of the theoretical value γ_c. Therefore, the guidelines in GB 50011-2010 [18] for box column steel frames could be followed, and the effect of panel zone shear deformation on inter-story displacements was not considered. This proves that a box-strengthened panel zone is more robust than the standard panel-zone doubler plates because it can form the box-shape panel zone in SMFs that use I-section columns.

4. Numerical Analysis Validation

4.1. Numerical Model Development

Based on the test, the cyclic loading analysis of the steel frame was carried out using ANSYS 16.0. The idealized bi-linear stress–strain steel constitutive model without strength degradation was used for the model. The yield stress σ_y, the yield strain ε_y, and the ultimate tensile stress σ_u and strain ε_u were determined through the carbon steel tensile test, as shown in Table 4 [22]. The bonded contact type imposed the beam-to-column interaction. Its beam and column components, skin plate, and column diaphragms all used SOLID95 elements. Since the overall deformation of the steel frame was investigated, the bolts, welds, and component's residual stresses were ignored in the modeling.

Table 4. Steel material properties used in ANSYS.

Steel Profiles	Thickness (mm)	Young's Modulus (MPa)	True Plastic Strain	True Stress (MPa)
			0.0	285
Column flange	15		0.2	525
			0.4	553
		2.02×10^5	0.6	567
Skin plate	16		0.8	580
			1.0	592
			0.0	307
Beam flange	11		0.2	534
			0.4	572
		2.04×10^5	0.6	600
Column web	10		0.8	619
			1.0	635
			0.0	332
			0.2	553
			0.4	607
Beam web	7	2.06×10^5	0.6	647
			0.8	670
			1.0	690

In addition, the point element MASS21 element, with six degrees of freedom, was chosen for the load points. Like the testing protocol, the boundary condition at the bottom of the column was assumed to limit all degrees of freedom, which means it was fixed. The lateral bracing constraints in the Z direction (vertical paper orientation) were imposed on the position of 1000 mm distanced from beam ends (see Figure 10a). The vertical loading point at the top beam end was subjected to the lateral displacement in the X-direction, and the repeated X-direction displacement was applied at the horizontal loading point of the beam-to-column joint, as shown in Figure 10a. The FE mesh used hexahedral elements with the most regular shapes for mapping and division. The box-strengthened panel zone within the height of the skin plate and the beam ends within the influence range of the joint measured 30mm; the other parts were 50 mm.

4.2. Numerical Analysis Verification

In general, ANSYS analysis revealed that the specimen's failure mode was the plastic hinge zone created in the beam end and panel zones maintained in an elastic condition, as well as buckling of the column flange at the column foot (Figure 10b). The overall failure pattern obtained from the numerical analysis was similar to the experimental one. Since many idealized factors existed in the numerical analysis model, the hysteretic curve derived by numerical analysis, as illustrated in Figure 9a, should have been fuller than the experimental curve. However, the two agreed on the backbone curve, which was used to calculate the critical parameters. Figure 9b shows that the test and ANSYS agreed on the initial stiffness and deformation trend. The ANSYS gave a maximum positive load of

663.9 kN and a displacement of 41.1 4 mm, which were 4% and 8.5% greater than the test result of 638.1 kN (37.9 mm), respectively. ANSYS produced a maximum negative load of −670.8 kN and a displacement of −45.5 mm, which were 0.002% and 0.2% greater than the test result of −669.7 kN (44.7 mm), respectively. Table 3 shows the numerically calculated inter-story displacement, Δ_n. Its inaccuracy from experimental measurements Δ_2 ranged from 4.2% to 5.8%. In conclusion, ANSYS numerical analysis provided reasonable accuracy and reliability and may be utilized for future parameter investigations.

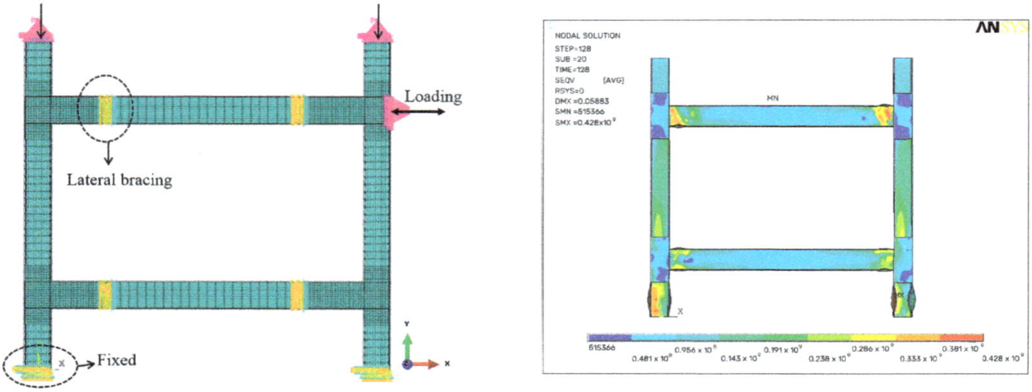

(a) (b)

Figure 10. FE analysis model and failure mode of the specimen: (**a**) FE model; (**b**) FE failure mode.

5. The Factors Influencing the Effect of Panel Zone Shear Deformation on Story Drift

There were two factors analyzed: one was the axial compression ratio (u), and the other was the beam-to-column linear stiffness ratio (sr).

5.1. The Influence of Axial Compression Ratio

For this NC series of specimens, the column, beam, steel, etc., were consistent with the specimens in the test; we only changed the axial compression ratio (u) from 0.15 to 0.75. The shear deformation angle γ of the panel zone and its proportion of Δ_s story drift under different axial compression ratios are illustrated in Figure 11. Δ_s was calculated by the following.

$$\Delta_s = \frac{\Delta_2 - \Delta_1}{\Delta_3} \times 100\%, \tag{5}$$

where Δ_1 = the approximate frame inter-story displacement without considering the shear deformation of the panel zone in. (mm), Δ_2 represents the frame inter-story displacement considering the shear deformation of the panel zone in. (mm), and Δ_3 is the frame inter-story displacement in. (mm) according to GB 50011-2010 [18].

From Figure 11a, it can be seen that the γ increased following the increase of loading displacement and axial compression ratio. However, as shown in Figure 11b, the proportion of Δ_s increased following the increase of axial compression ratio but declined following the increase of loading displacement, and the proportions of Δ_s to the story drift were from about 10% to 4%. Consequently, the box-strengthened panel zone connection reduced the development of panel zone shear deformation and brittle failure in the beam end weld zone, and the story drift increment of the steel frame did not exceed 10%. When considering the design of real SMFs, the axial compression ratio is generally strictly controlled and will not exceed 0.75. At this time, the shear deformation of the panel zone itself contributes less than 10% to the inter-story displacement. Therefore, its impact can be ignored during design when the axial compression ratio is less than 0.75.

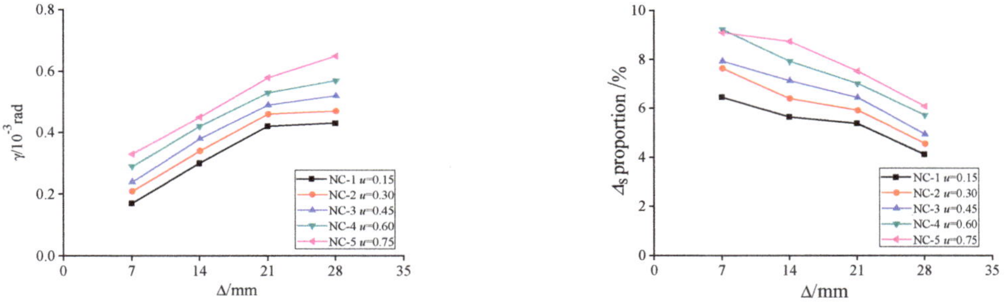

(a) (b)

Figure 11. Shear deformation comparison of NC series of specimens: (**a**) shear deformation angle γ; (**b**) proportion of Δ_S.

5.2. The Influence of Beam-to-Column Stiffness Ratio

In the Chinese standard GB50011-2010 [18], strong columns and weak beams are provisions that must followed. The most essential condition affecting the strong columns and weak beams is the linear stiffness ratio of beams to columns. Therefore, the primary way to change sr (beam-to-column stiffness ratio) for SRC series specimens is to change the beam length (frame span). The column was HW 300 × 300 × 10 × 15, and the beam was HN 350 × 175 × 7 × 11. We changed the frame span from 3.9 m to 3.3 m, 3.0 m, 2.7 m, and 2.6 m, and the linear stiffness ratios corresponded to 0.66, 0.78, 0.858, 0.953, and 1.0.

In Figure 12a, the γ increases following the increasing loading displacement and linear stiffness ratio. However, from Figure 12b, it can be seen that the proportion of Δ_S increases following the increase of linear stiffness ratio, but the proportion of Δ_S declines following the increase of loading displacement. However, the proportions of Δ_S to the story drift were between 10% and 5%. Therefore, when the linear stiffness ratio of beam-to-columns did not exceed 1.0, the shear deformation of the panel zone itself contributed less than 10% to the inter-story displacement, and its impact can therefore be ignored during design.

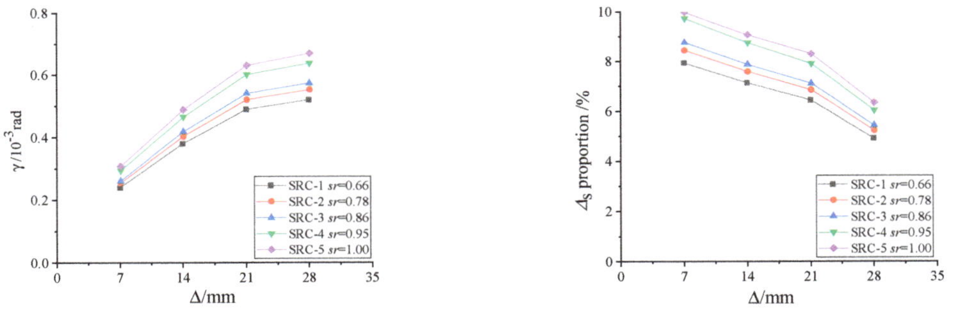

(a) (b)

Figure 12. Shear deformation comparison of SRC series of specimens: (**a**) shear deformation angle γ; (**b**) proportion of Δ_S.

6. Conclusions

This paper investigated the influence of a box-strengthened panel zone on the seismic performance and story drift of steel SMFs by test and numerical methods. The following conclusions were obtained.

1. The seismic performance of steel frame specimens using the box-strengthened panel zone is good.

2. The box-strengthened panel zone is a type of strong panel zone. It is more robust than the standard panel-zone doubler plates and can effectively reduce the inter-story drift of steel SMFs.
3. The axial compression ratio and the beam-to-column linear stiffness ratio have the same influence on the shear deformation of the panel zone.
4. Under design conditions where the axial compression ratio is less than 0.75 and the beam-to-column linear stiffness ratio is less than 1.0, the contribution of panel zone shear deformation to the frame's inter-story drift does not exceed 10%. Its impact can be ignored during design.

Author Contributions: Conceptualization, L.L.; software, J.Z; validation, J.Z., G.Z., H.P. and B.L.; resources, L.L.; writing—original draft preparation, L.L.; writing—review and editing, L.L. and H.H.; project administration, L.L.; funding acquisition, L.L. All authors have read and agreed to the published version of the manuscript.

Funding: This research was funded by the Nature Science Foundation of China (NSFC), grant number 51278061.

Data Availability Statement: The data presented in this study are available on request from the corresponding author. The data are not publicly available due to patent protection.

Acknowledgments: The authors thank the China Scholarship Council.

Conflicts of Interest: The authors declare no conflict of interest.

References

1. León, D.D.; Reyes, A.; Yu, C. Probabilistic assessment of connections for steel buildings on seismic zones. *J. Constr. Steel Res.* **2013**, *88*, 15–20. [CrossRef]
2. Memari, M.; Mahmoud, H. Performance of steel moment resisting frames with RBS connections under fire loading. *Eng. Struct.* **2014**, *75*, 126–138. [CrossRef]
3. Wu, H.H.; Zhou, T.H.; Liao, F.F.; Lv, J. Seismic behavior of steel frames with replaceable reinforced concrete wall panels. *Steel Compos. Struct.* **2016**, *22*, 1055–1071. [CrossRef]
4. Morshedi, M.A.; Dolatshahi, K.M.; Maleki, S. Double reduced beam section connection. *J. Constr. Steel Res.* **2017**, *138*, 283–297 [CrossRef]
5. Lu, L.; Wang, D.; Ding, K.; Yan, H.; Hao, H. A Proposal for a Simple Method for Determining the Concrete SlabWidth of Composite Beam-to-Column Joints. *Appl. Sci.* **2021**, *11*, 9613. [CrossRef]
6. Koetaka, Y.; Chusilp, P.; Zhang, Z.; Ando, M.; Suitad, K.; Inoue, K.; Uno, N. Mechanical property of beam-to-column moment connection with hysteretic dampers for column weak axis. *Eng. Struct.* **2005**, *27*, 109–117. [CrossRef]
7. Cabrero, J.M.; Bayo, E. The semi-rigid behavior of three-dimensional steel beam-to-column joints subjected to proportional loading. Part I: Experimental evaluation. *J. Struct. Eng.* **2007**, *63*, 1241–1253.
8. Cabrero, J.M.; Bayo, E. The semi-rigid behavior of three-dimensional steel beam-to-column joints subjected to proportional loading. Part II: Theoretical model and validation. *J. Struct. Eng.* **2007**, *63*, 1254–1267.
9. Kim, S.D.; Kim, S.S.; Ju, Y.K. Strength evaluation of beam-column connection in the weak axis of H-shaped column. *Eng. Struct.* **2008**, *30*, 1699–1710. [CrossRef]
10. Lee, E.T.; Kang, M.J.; Kim, S.B.; Kim, S.S. Experimental study on structural performance of the new shaped weak-axis connection in full-scale test. *Int. J. Steel Struct.* **2016**, *16*, 685–696. [CrossRef]
11. Lee, K.; Li, R.; Jung, H.; Chen, L.; Oh, K. Cyclic testing of weak-axis steel moment connections. *Steel Compos. Struct.* **2013**, *15*, 507–518. [CrossRef]
12. Lee, K.; Li, R.; Jung, H.; Chen, L.; Oh, K.; Kim, K.S. Cyclic testing of steel column-tree moment connections with various beam splice lengths. *Steel Compos. Struct.* **2014**, *16*, 221–231. [CrossRef]
13. Kozlowski, A. Component method model for predicting the moment resistance, stiffness and rotation capacity of minor axis composite seat and web site plate joints. *Steel Compos. Struct.* **2016**, *20*, 469–486. [CrossRef]
14. Shim, H.J.; Lee, E.T.; Kim, S.B.; Kim, S.S. Development and performance evaluation of weak-axis column bending connections for advanced constructability. *Int. J. Steel Struct.* **2014**, *14*, 369–380. [CrossRef]
15. Kim, Y.J.; Oh, S.H.; Moon, T.S. Seismic behavior and retrofit of steel moment connections considering slab effects. *Eng. Struct.* **2004**, *26*, 1993–2005. [CrossRef]
16. Adlparvar, M.R.; Vetr, M.G.; Ghaffari, F. The importance of panel zone shear strength on seismic behavior of improved slotted-web beam connections. *J. Struct. Eng.* **2017**, *17*, 307–318. [CrossRef]
17. Mahmoud, M.; Morteza, N.; Amir, K. Panel zone rigidity effects on special steel moment-resisting frames according to the performance based design. *World Acad. Sci. Eng. Technol.* **2009**, *50*, 925–931.

18. *GB50011-2010*; Code for Seismic Design of Buildings. China Architecture & Building Press: Beijing, China, 2010.
19. Lu, L.F.; Xu, Y.L.; Lim, J.B.P. Mechanical performance of a new I-section weak-axis column bending connection. *Steel Compos. Struct* **2018**, *26*, 31–44.
20. *ANSI/AISC 341-22*; Seismic Provisions for Structural Steel Buildings. American Institute of Steel Construction: Chicago, IL, USA, 2022.
21. FEMA 273 NEHRP. *Commentary on the Guidelines for the Seismic Rehabilitation of Buildings*; Federal Emergency Management Agency: Washington, DC, USA, 1997.
22. Lu, L.F.; Xu, Y.L.; Zheng, H. Investigation of composite action on seismic performance of weak-axis column bending connections. *J. Constr. Steel Res.* **2017**, *128*, 286–300. [CrossRef]

buildings

MDPI

Article

An Investigation on the Seismic Behavior of the End-Plate Connection between a Steel Beam and the Weak-Axis of an H-Shaped Column Using a U-Shaped Connector

Shaofeng Nie [1,*], Siyuan Chen [1], Weijie Wu [1], Xin Jiang [1], Shuo Wang [2], Yongzhen Lin [1] and Mengxue Liu [1]

1 School of Civil Engineering, Chang'an University, 75 Chang'an Middle Road, Xi'an 710061, China; 2022228025@chd.edu.cn (S.C.); 2022128039@chd.edu.cn (W.W.); 2022128022@chd.edu.cn (X.J.); 2022228033@chd.edu.cn (Y.L.); 2023128028@chd.edu.cn (M.L.)

2 School of Civil Engineering, Huxi Vocational and Technical College, Xuzhou 221600, China; wsvermouth_ga@126.com

* Correspondence: niesf@chd.edu.cn

Abstract: The seismic behavior of the end-plate connections between a steel beam and the weak axis of the H-shaped steel column using a U-shaped connector was investigated using numerical analysis. Finite element (FE) models were established using ABAQUS 6.14 software, and the applicability of the modeling approach was verified by comparing the numerical results with the relevant experimental results. This parametric study of the joint was carried out to analyze the effects of the thickness of the U-shaped connectors, the thickness of the end-plates, the axial compression ratio of the columns, and the linear stiffness ratio of the beam to the column. The results show that the U-shaped connector set in the weak axis of the H-shaped column can form a box-shaped panel zone with the column flange and web. The volume of the panel zone and its resistance to shear deformation are increased through this connection. Finally, the recommended reasonable ranges for the thickness of the U-shaped connector, the thickness of the end-plates, the axial compression ratio of the columns, and the linear stiffness ratio of the beam to the column are proposed in this paper.

Keywords: weak-axis beam-to-column connection; U-shaped connector; end-plate connection; seismic behavior; failure mode; numerical analysis

1. Introduction

The semi-rigid connection behavior is between the rigid and hinge connections, which can effectively transfer the shear force and bending moment. A relative rotational angle will exist between the beam and the column, which has a certain rotational ability [1]. As a classic semi-rigid connection, the high-strength bolted end-plate connection has excellent ductility and energy dissipation capacity [2,3], which can reduce damage in an earthquake.

There is much research on the high-strength bolted end-plate connection behavior in the strong-axis direction of H-shaped steel columns. Shi et al. [4] applied cyclic loads to eight end-plate connections with different structures for experimental investigation. The results showed that excellent ductility and energy dissipation capacity were applied to the semi-rigid end-plate connection of beam to column and could be applied in multi-story steel frame structures. Shi et al. [5] used the finite element (FE) method to investigate the mechanical behavior of two types of semi-rigid connections: flush end-plate joints and extended end-plate joints. The results showed that setting the stiffener on the extended end-plate could give the joint better mechanical behavior. Zhao [6] conducted a series of experimental studies on the stiffness of the extended end-plate strong-axis connection and obtained the effect of the end-plate on the stiffness of the strong-axis joint. Ghobarah et al. [7] conducted an experimental investigation on the panel zone mechanical behavior of the strong-axis end-plate connection. The results showed that the panel zone was an important member of structural energy dissipation, ductility, and bearing capacity stability. Sumner et al. [8] performed a quasi-static loading test on the extended beam-to-column end-plate connection with end-plate stiffener. The test results showed that the joint's strength, stiffness, and ductility could be improved effectively by adding the end-plate stiffener.

In the weak-axis connection of the H-shaped steel column, the steel beam is connected by a wedge-shaped stiffened plate in the traditional weak-axis direction. However, the plastic hinge formation position of the beam end is generally near the butt welding of the beam, making it easy to cause welding damage. Cabrero et al. [9] proposed a joint with two separate additional plates welded to the beam flange as an end-plate in the weak-axis direction and carried out a monotonic loading test. The results showed that the bearing capacity and rotation stiffness of the weak axis of the column were improved by increasing the thickness of the end-plate. Nie et al. [10] carried out seismic research on the end-plate edge connection of a steel beam to the weak axis of an H-shaped column with a U-shaped connector and proposed the end-plate's thickness and the stiffener's structural value. Lu and Wang [11–13] proposed the box-shaped panel zone end-plate connection for the weak-axis column, and the seismic performance of the joints with stiffener at the beam end was analyzed. The results showed that the stiffness and bearing capacity of the joint could be improved effectively by setting a stiffener in the joint. Tagawa et al. [14] carried out experimental research and numerical analysis on the mechanical behavior of bolted end-plate connections with channels and discussed the mechanical behavior of the panel zone. The results showed that the joint's bearing capacity and shear capacity could be improved by setting a stiffener in the channel.

Conventional weak-axis connections use brackets that require four connections, complex construction, and inconvenient transportation [15]. Kim [16] conducted the experimental and theoretical analysis of the beam–column connection of the H-shaped column's weak axis, which was used to connect the steel beam and column web. The connections were designed with T-shaped connectors, specifically split-tee, welded split-tee, end-plate, and conventional bracket. These T-shaped connectors and end-plates were connected directly to the column web, which would induce large deformation and lower stiffness of the column web. The results showed that the tensile-connected bolt determines the maximum strength and ductility of the connection. Shim [17] proposed a beam-to-column connection, in which an H-shaped steel beam is joined to an H-shaped steel column by bolted splices at the top flange without a scallop at the web.

For other reinforcing methods on joints, Nassiraei et al. [18] evaluated the effect of the ring, the plate, and the joint geometry on the failure mechanisms of tubular X joints subjected to axial compression in different fire conditions. The results showed that both the external ring and plate could increase the joints' initial stiffness and ultimate capacity in fire conditions. Rahgpzar et al. [19] adopted the Fiber-Reinforced Plastic (FRP) plate to improve the flexural strengthening of precast beam–column joints. Experimentation and finite element analyses led to the conclusion that longitudinal beam reinforcement ratios directly influenced the load-bearing capacity. Nie et al. [20] proposed a weak-axis connection form of a steel tube-reinforced beam with an opening in the web. The results showed that the opening distance greatly affected the position of the plastic hinge. The numerical analysis of the seismic behavior of square steel tube confined steel-reinforced concrete column-RC ring beam joint was conducted by Nie et al. [21]. The results showed that the ring beam played a good role in protecting the core area of the joint.

The stiffness and strength of conventional weak-axis connections using brackets or other T-shaped connectors were lower than those of the strong-axis connections. The traditional connection method weakened the column and decreased the stiffness of the panel zone in the weak-axis direction. Therefore, meeting the seismic needs of the 'strong joint with weak member' is problematic. This paper proposes the end-plate connection of a steel beam to the weak axis of an H-shaped column using a U-shaped connector. A strong box-shaped panel zone of this kind of joint is formed in the weak-axis direction so that the joint stiffnesses in the strong-axis and weak-axis directions are proximal. It is easy to satisfy the anti-seismic demand of 'strong joint, weak members'. In addition, the U-shaped connector and end-plate could be connected easily with high-strength bolts. The assembly efficiency, ductility, and energy dissipation capacity of the panel zone could be improved.

This paper used ABAQUS 6.14 software to conduct in-depth research on the seismic behavior of the novel end-plate joint. The effect of the thickness of the U-shaped connector, the stiffener's length and height, the beam end-plate's thickness on the failure mode, and the seismic behavior of the joint are analyzed in detail, and design methods are proposed.

2. Design of Specimen

The BASIC specimen was designed according to the 'Standard for Design of Steel Structures' (GB 50017-2017) [22] and references [23,24]. The stiffener length of the beam was 90 mm, the height was 40 mm, and the thickness was 10 mm. The M24 high-strength bolts were grade 10.9, and the steel grade was Q235B. The pre-tension force of 225 kN was applied to the high-strength bolt. The configuration and dimensions of the BASIC specimen are shown in Figure 1.

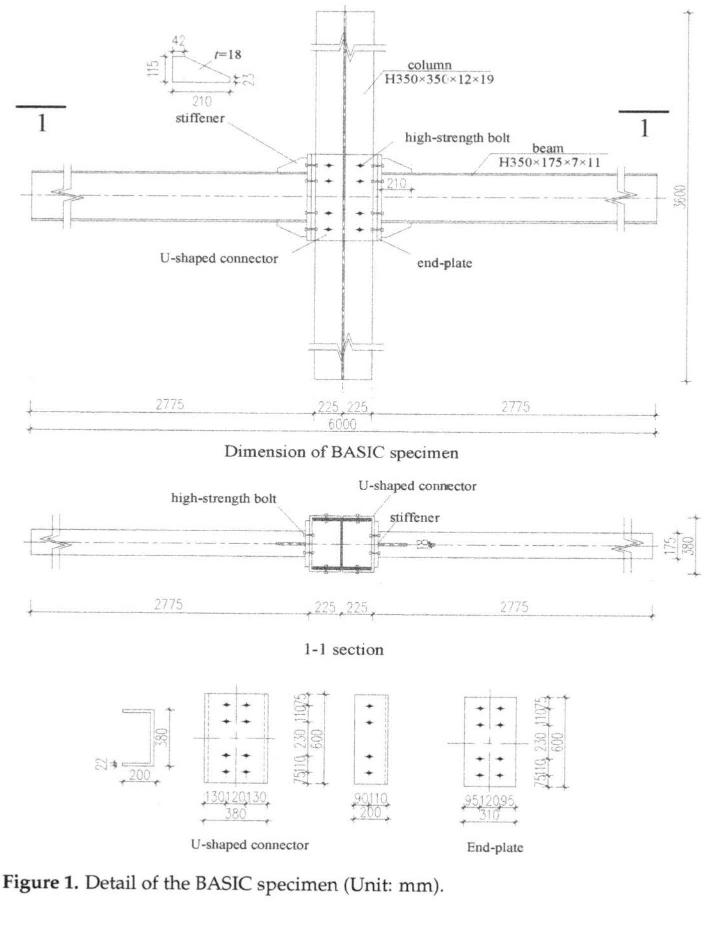

Figure 1. Detail of the BASIC specimen (Unit: mm).

3. Establishment and Verification of Finite Element Model

3.1. Constitutive Relationship Setting and Element Selection

A double-broken line model simulated the stress–strain relationships of steel and high-strength bolts. The elastic modulus of steel was 206,000 MPa, and the Poisson's ratio was 0.3. The yielding strength, the ultimate strength, and the ultimate strain of the steel were 235 MPa, 420 MPa, and 0.15, respectively. The yielding strength, the ultimate strength, and the ultimate strain of the high-strength bolts were 940 MPa, 1130 MPa, and 0.10, respectively.

The solid elements (C3D8R) were used for all parts of the finite element (FE) model. A mesh-sensitivity study was conducted in the FEM analysis for the specimens to investigate the effect of the mesh size on the behavior of the specimens. By calculating and comparing different mesh sizes for the specimen, the model adopted the following sizes: 40 mm × 40 mm for steel beam and column, 10 mm × 10 mm for U-shaped connector, and 6 mm × 6 mm for high-strength bolts. The mesh at the panel zone and the bolt hole were encrypted.

3.2. Boundary Conditions and İnteractions

For the BASIC specimen, the beam-to-column connection panel zone grids were encrypted. Reference points RP1 and RP3 were established at the top and bottom of the column, respectively. All the freedom degrees on the top and bottom surfaces of the column were coupled with RP1 and RP3, respectively. In addition, the freedom degrees on the side flange surface of the top column with a length of 50 mm were coupled with RP2. The displacement constraints were imposed on the side surface of the beam flange within 50 mm from the loading point to prevent the instability of the lateral displacement of the model during the loading process. The loadings were applied to the corresponding reference points to make sure the loading conditions and constraints of the BASIC specimen were consistent with the experimental conditions.

Regarding the interaction setting, the tangential contact was defined as Coulomb friction, and the value of the frictional coefficient was 0.45 [22]. In addition, the normal direction was set as hard contact. The U-shaped connector was connected to the steel column by the 'Tie' command in ABAQUS, and the 'Tie' interaction was also used to connect the beam to the end-plate. The boundary conditions and grid division are shown in Figure 2.

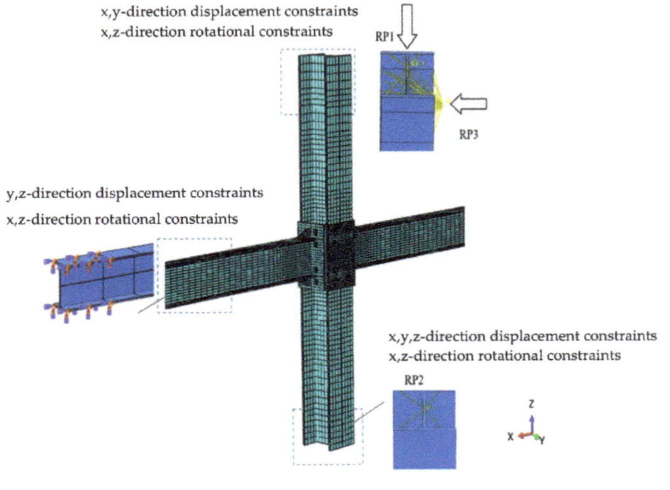

Figure 2. Boundary conditions and grid division of the BASIC specimen.

3.3. Loading Process

The loading scheme and loading process are illustrated in Figure 3, which were carried out by following the reference [25].

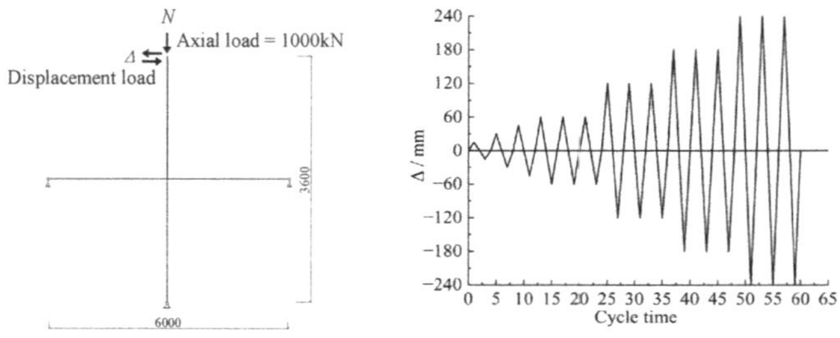

(**a**) (**b**)

Figure 3. Loading method: (**a**) loading scheme (unit: mm); (**b**) loading process.

The constant axial compression of 1000 kN was applied on the top of the column first, and then the cyclic loading was applied on the point RP2. Firstly, a monotonic load was applied to the loading point, and then the yielding load was defined according to the load-to-displacement curve. The yielding displacement corresponding to the yielding load was 60 mm. In the cyclic loading process, 20% of the yielding displacement was taken for displacement loading before the yield of the specimen with one cycle. After yielding, the loading displacement was increased by 60 mm per stage, using a three-time cycle. The loading was stopped after the load decreased by approximately 85% of the ultimate load or the specimen was destroyed.

3.4. Verification of FEM

A similar joint about the weak-axis column with a box-shaped panel zone was proposed by Lu [12,13]. The S-4 specimen in the test was selected just for FEM validation. The finite element model analysis (FEM analysis) results were compared with the experimental results. The details of the S-4 specimen are shown in Figure 4.

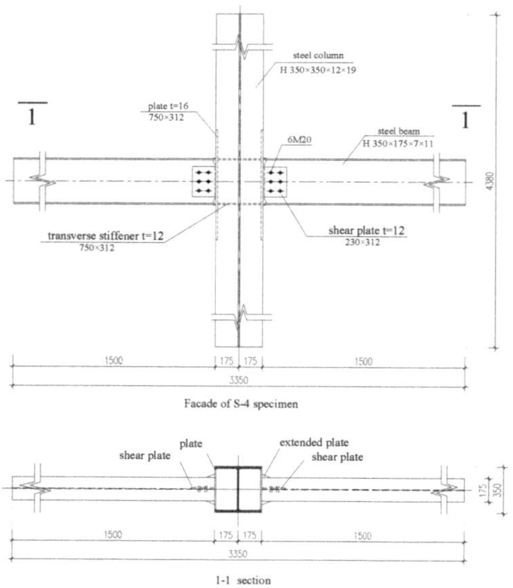

Figure 4. Details of the S-4 specimen (unit: mm).

The comparison of the failure mode, moment-rotation hysteresis curves, and skeleton curves obtained by FEM analysis and experiment are shown in Figures 5 and 6. The comparative studies of the results are illustrated in Table 1.

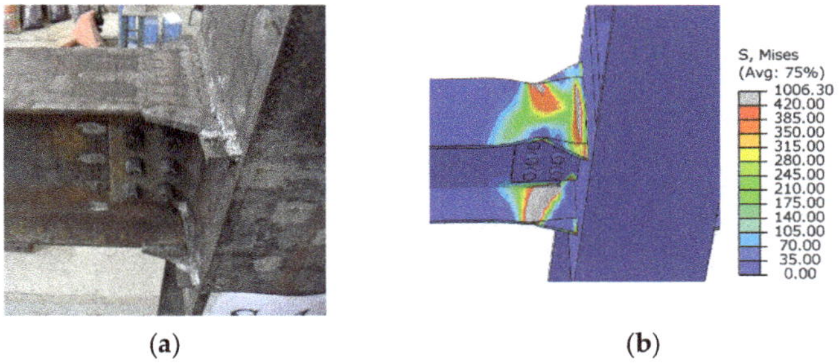

Figure 5. Comparison of failure mode: (**a**) failure mode in the test result; (**b**) failure mode in the FEM analysis result.

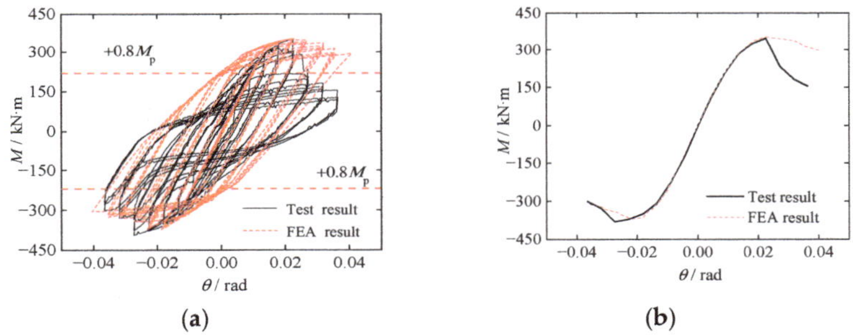

Figure 6. Comparison of hysteric curves and skeleton curves: (**a**) hysteresis curves of the test result and FEM analysis result; (**b**) skeleton curves of the test result and FEM analysis result.

Table 1. Comparison of FEM analysis and test results.

Specimen	Loading Direction	M_y/kN·m	θ_y/rad	M_{max}/kN·m	θ_{max}/rad	M_u/kN·m	θ_u/rad	μ
Test result	+	289.5	0.014	347.2	0.021	295.1	0.025	1.74
	−	311.7	0.014	393.7	0.021	334.7	0.032	2.29
FEM analysis results	+	305.4	0.015	354.0	0.021	295.3	0.041	2.86
	−	312.5	0.015	371.1	0.022	304.7	0.041	2.87

Note: M_y, M_{max}, and M_u represent the yielding moment, the maximum moment, and the ultimate moment of the joint. θ_y, θ_{max}, and θ_u represent the inter-story drift under the yielding moment, the maximum moment, and the ultimate moment of the joint, respectively. μ represents the rotational ductility coefficient, $\mu = \theta_u/\theta_y$.

The hysteresis curves of moment vs. rotation are used to investigate the joint's rotational capacity and seismic behavior. The inter-story drift θ and moment of the beam end M of the joint are calculated by Equations (1) and (2), respectively [26,27]. When the joint reached 0.04 rad inter-story drift, the joint bearing capacity should be above $0.8M_p$ [28].

$$\theta = \arctan\frac{\Delta}{H}, \tag{1}$$

$$M = R_1 l, \tag{2}$$

where Δ represents the horizontal displacement; H represents the height from the loading point to the bottom of the column; R_1 represents the reaction force of the end of the left beam; l represents the distance from the beam end to the surface of the U-shaped connector; M_p represents the plastic moment of the whole section of the beam.

The comparisons of failure mode between the test and FEM analysis are shown in Figure 5. During the test, severe cracks appeared in the welding near the connection between the beam and the cover-skin plate. In the FEM analysis, the high-stress region was concentrated near the welding between the beam and the cover-skin plate, where the stress values reached the ultimate stress.

The comparisons of hysteric curves and skeleton curves between the test and FEM analysis are shown in Figure 6. The hysteretic curves obtained from the FEM analysis and the test had the same law. The test curve pinch was mainly due to the significant relief to the beam flange weld on the other side after the cracking of one side of the beam flange welding. While the model in the FEM was idealized, there was no such pinch phenomenon in the hysteretic loops. The load versus displacement curves of the test showed an apparent drop, while the curves were smoother in the FEM. This is attributed to the fact that the beam flange welds cracking occurred in the test, which was hard to simulate in the FEM because the welding damage was not considered to overcome the convergence problem. The skeleton curves obtained from the FEM analysis are well in agreement with those obtained from the experiment. The stiffness of the curve obtained from the test is the same as the stiffness obtained from the FEM. The difference between the test result and FEM results of M_{max} was less than 2%, which validated the correctness of the FEM.

4. Seismic Behavior of BASIC and UR Specimens

The UR specimen without end-plate stiffeners was established. The rest of the parameters of the UR specimen were the same as those of the BASIC specimen. In the FEM analysis, when the load dropped to 85% of the maximum load, the specimen was considered destroyed. In terms of data processing, the secant stiffness was used to represent the stiffness degradation of the joint [25], and the general yielding moment method was used to determine the yielding moment and yielding rotation [29].

4.1. Moment-Rotation Hysteresis Curve and Skeleton Curve

The moment–rotation hysteresis and skeleton curves of the BASIC and UR specimens are shown in Figure 7, respectively. The stiffness of the joint within the elastic phase is defined as the initial stiffness R_{ki}, which is calculated by Equation (3). The comparison of FEM analysis results of UR and BASIC specimens under cyclic loading is illustrated in Table 2.

$$R_{ki} = \left.\frac{dM}{d\theta}\right|_{\theta=0}, \tag{3}$$

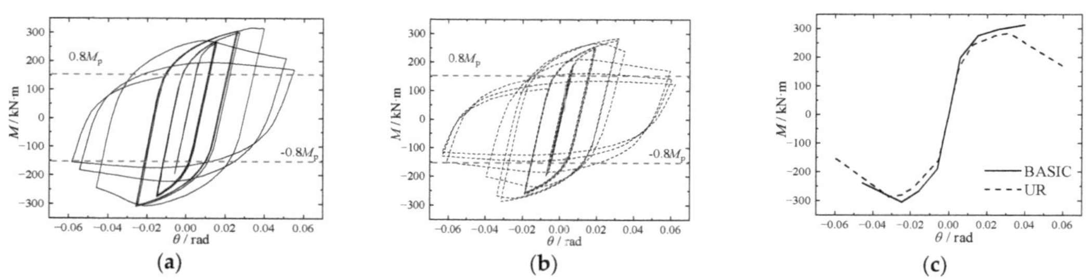

Figure 7. Comparison of hysteresis curves and skeleton curves of specimens: (**a**) hysteresis curve of BASIC specimen; (**b**) hysteresis curve of UR specimen; (**c**) skeleton curves of BASIC and UR specimens.

Table 2. Comparison of FEM analysis results.

Specimen	M_y/kN·m	θ_y/rad	M_{max}/kN·m	θ_{max}/rad	M_u/kN·m	θ_u/rad	μ	R_{ki}
UR	230.3	0.012	286.5	0.032	170.3	0.050	4.2	31,362
BASIC	236.5	0.013	316.8	0.033	264.9	0.046	3.6	33,048

As shown in Figure 7, the curves of the BASIC and UR specimens were relatively full. At the initial stage of loading, the skeleton curves of the two specimens coincide with each other. The moment of the two specimens decreased slowly after reaching the maximum value.

As indicated in Table 2, the yielding moment M_y, the maximum moment M_{max}, and the initial stiffness of the joint R_{ki} of the BASIC specimen were 2.7%, 10.6%, and 5.4% higher than those of the UR specimen, respectively. The stiffeners were set between the beam and the U-shaped connector, which could improve the seismic behavior of the specimen.

4.2. Comparative Analysis of Failure Mode

Failure modes of the UR specimen and the BASIC specimen are shown in Figures 8 and 9.

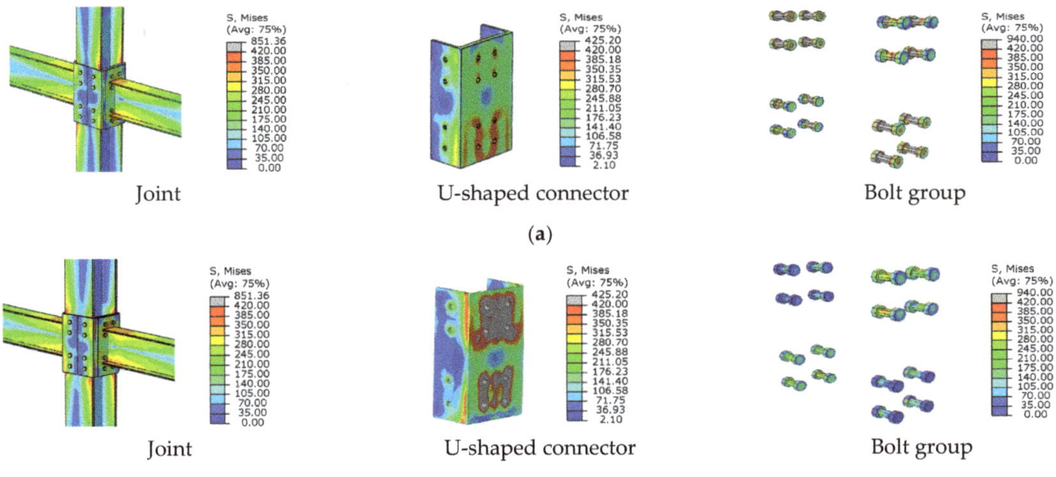

Figure 8. The contour of the Von Mises stress of the UR specimen: (**a**) yielding load; (**b**) failure load.

For the UR specimen, the yielding and failure modes were significantly different from the BASIC specimens. In the yielding stage, the stress in some regions of the U-shaped connector flange surpassed the ultimate stress value, and the failure at the bolt hole on the web was more serious. In the failure stage, the ultimate stress at the bolt holes of the U-shaped connector spread to most of the web region. In the process of cyclic loading, the stress of the high-strength bolts was lower than their ultimate capacity, indicating that the high-strength bolts of this type of joint satisfied the bearing capacity requirements. The failure position of the UR specimen was in the panel zone, which did not satisfy the anti-seismic demand of 'strong joint, weak members.

For the BASIC specimen, the two outermost bolt holes on the U-shaped connector web had the highest stresses at the yielding load. At the failure load, although the stress in some regions of the U-shaped connector web reached the ultimate stress, the whole seismic behavior of the specimen was excellent. In addition, the maximum stress at the beam was near the lateral region of the stiffener, so the beam produced obvious buckling deformation on the outer side of the stiffener. In summary, the final failure mode of the BASIC specimen was that the plastic hinge was formed far away from the end-plate of the beam, which could better protect the panel zone, improve the seismic behavior of the U-shaped connector, and avoid the welding fracture which caused by the stress concentration of the beam and the end-plate welding.

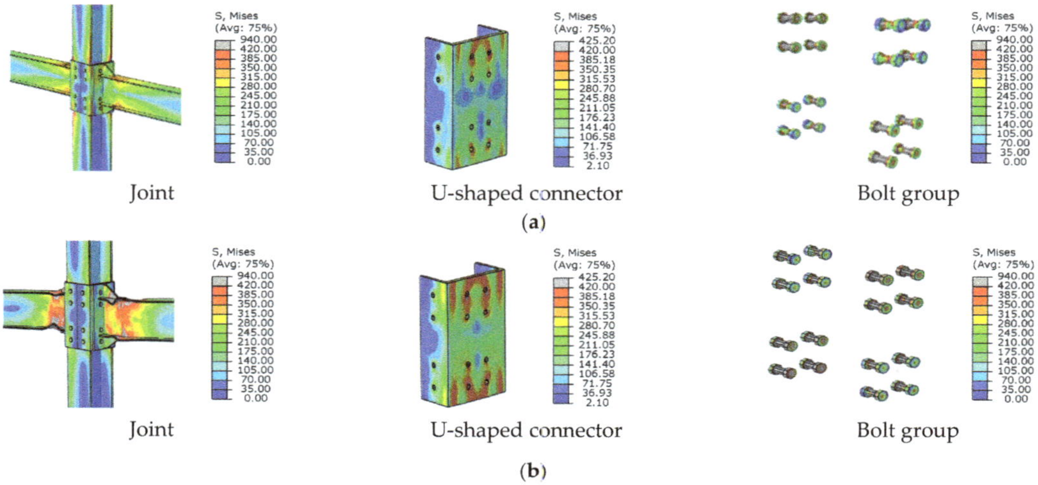

Figure 9. The contour of the Von Mises stress of the BASIC specimen: (**a**) yielding load; (**b**) failure load.

5. Parametric Study

The influence of the thickness of the U-shaped connector, the end-plate thickness, the axial compression ratio of the column, and the linear stiffness ratio of the beam to the column on the seismic behavior of the joint were analyzed. The parameters and results of specimens are illustrated in Table 3.

Table 3. The parameters and results of specimens.

Specimen	A/mm	B/mm	C	D	M_y/kN·m	θ_y/rad	M_{max}/kN·m	θ_{max}/rad	M_u/kN·m	θ_u/rad	μ
BASIC	22	24	0.25	0.60	236.5	0.013	316.8	0.033	264.9	0.047	3.62
UT-1	18	24	0.25	0.60	232.2	0.013	297.5	0.034	256.3	0.049	3.91
UT-2	20	24	0.25	0.60	234.4	0.013	309.3	0.028	261.4	0.050	3.84
UT-3	24	24	0.25	0.60	241.4	0.014	318.9	0.029	275.9	0.050	3.52
UT-4	26	24	0.25	0.60	245.4	0.014	323.3	0.029	224.3	0.048	3.37
DB-1	22	18	0.25	0.60	232.0	0.013	307.2	0.048	249.2	0.048	3.69
DB-2	22	22	0.25	0.60	235.6	0.013	316.3	0.047	316.3	0.047	3.61
DB-3	22	26	0.25	0.60	245.4	0.013	329.1	0.044	329.1	0.044	3.38
DB-4	22	30	0.25	0.60	247.4	0.013	330.2	0.041	330.2	0.042	3.15
ZY-1	22	24	0.30	0.60	233.2	0.013	313.3	0.026	262.5	0.047	3.58
ZY-2	22	24	0.40	0.60	230.7	0.013	312.3	0.029	261.3	0.045	3.47
ZY-3	22	24	0.60	0.60	224.9	0.013	310.2	0.030	258.3	0.045	3.45
ZY-4	22	24	0.80	0.60	223.2	0.013	308.9	0.026	256.4	0.044	3.31
BLS-1	22	24	0.25	0.32	228.3	0.013	312.8	0.040	262.5	0.047	3.58
BLS-2	22	24	0.25	1.05	242.5	0.013	323.0	0.035	281.2	0.045	3.42
BLS-3	22	24	0.25	1.49	244.5	0.013	324.2	0.035	284.6	0.039	2.97
BLS-4	22	24	0.25	2.10	245.6	0.014	347.2	0.035	290.1	0.041	2.95

Note: A represents the thickness of the U-shaped connector; B represents the thickness of the end-plate; C represents the axial compression ratio; D represents the linear stiffness ratio of the beam to column.

5.1. The Thickness of the U-Shaped Connector

The thickness of the U-shaped connector was varied with values of 18 mm (0.95t_f), 20 mm (1.0t_f), 24 mm (1.26t_f), and 26 mm (1.37t_f), and t_f represented the thickness of the column flange. UT-1, UT-2, UT-3 and UT-4 specimens were established, respectively. The contours of the Von Mises stress of UT series specimens under failure modes are shown in Figure 10, and hysteresis curves and skeleton curves are shown in Figures 11 and 12, respectively.

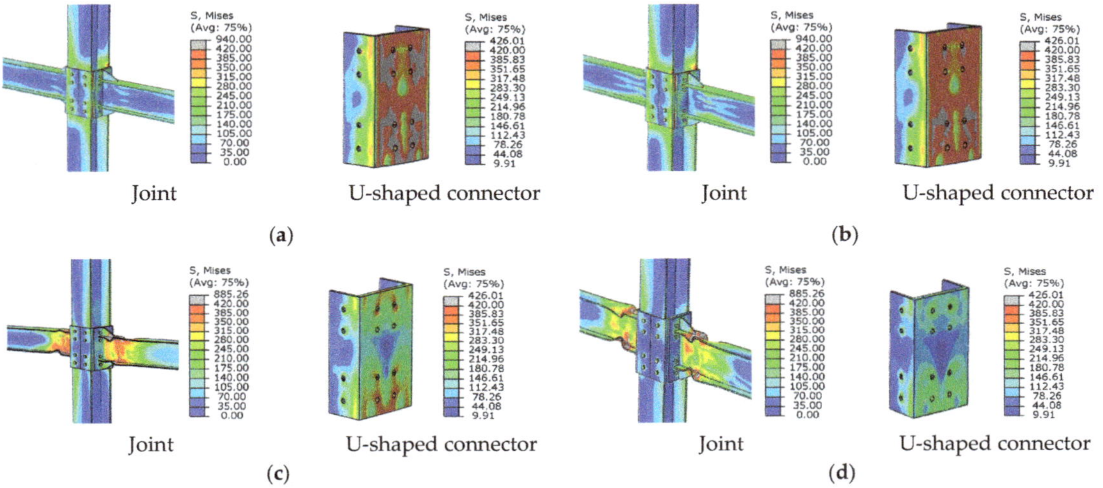

Figure 10. The contour of the Von Mises stress of UT specimens under failure modes: (**a**) UT-1; (**b**) UT-2; (**c**) UT-3; (**d**) UT-4.

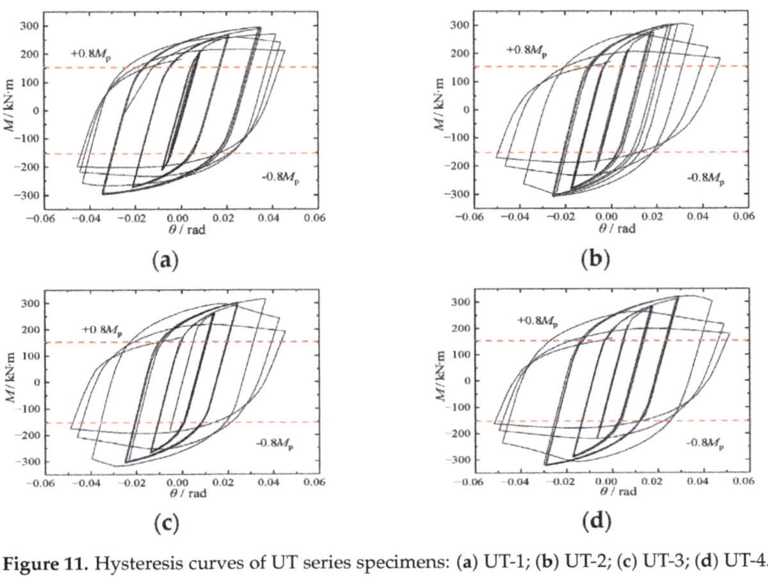

Figure 11. Hysteresis curves of UT series specimens: (**a**) UT-1; (**b**) UT-2; (**c**) UT-3; (**d**) UT-4.

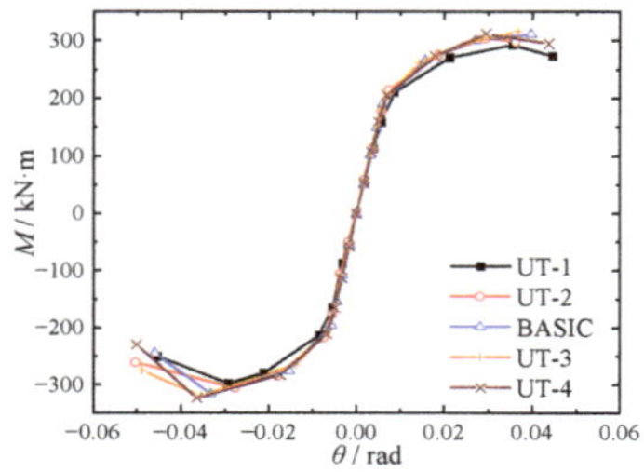

Figure 12. Skeleton curves of UT specimens.

It could be seen that the failure modes of UT-3 and UT-4 specimens were similar to the BASIC specimen. The buckling occurred at the beam flange and web far from the joint. The stress value of the U-shaped connector was low, and the joint deformation was small. For UT-1 and UT-2 specimens, the bearing capacity of the U-shaped connector was poor because of its thinner thickness and lower stiffness. A significant plastic deformation occurred in most regions of the U-shaped connector web, which caused the failure of the U-shaped connector first.

As shown in Figures 12 and 13, the hysteresis curves of UT series specimens were relatively full. At the initial load stage, the skeleton curves of each specimen were the same, and the maximum load increased slightly with the increase in the thickness of the U-shaped connector.

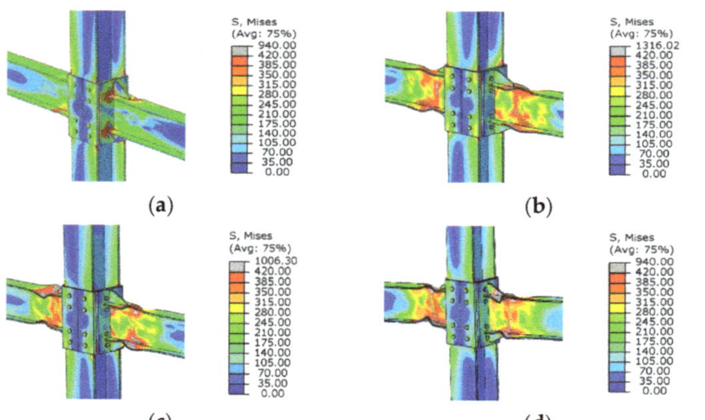

Figure 13. The contour of the Von Mises stress of DB specimens under failure modes: (**a**) DB-1; (**b**) DB-2; (**c**) DB-3; (**d**) DB-4.

When the thickness of the U-shaped connector was varied from 18 to 20 mm, 18 to 22 mm, 18 to 24 mm, and 18 to 26 mm, the maximum moment was increased by 4.0%, 6.5%, 7.2%, and 8.7%, respectively and the ductility coefficient of specimens was reduced by 5.6%, 11.7%, 12.5%, and 15.9%, respectively. It could be seen that the ductility and maximum moment of the beam end were significantly affected by the thickness of the U-shaped connector.

To ensure the plastic hinge forms within a reasonable range of the beam, t_u, the U-shaped connector's thickness should be in the range of $1.15t_f$ to $1.30t_f$.

5.2. The Thickness of the End-Plate

The end-plate thickness of the joint could be calculated according to the [29] by Equation (4). The calculated result showed that the end-plate thickness should not be less than 22 mm.

$$t_d \geq \sqrt{\frac{6e_f e_w N_t}{\left[e_w b + 2e_f\left(e_f + e_w\right)\right]f}}, \tag{4}$$

where N_t represented the design value of the tensile force of a high-strength bolt; e_w and e_f represented the distance from the center of the bolt to the web and the flange plate, respectively; b represented the width of the end-plate; f represented the design strength of the end-plate.

The end-plate thickness was varied, with values of 18 mm, 22 mm, 26 mm, and 30 mm. DB-1, DB-2, DB-3, and DB-4 specimens were established, respectively. The contours of the Von Mises stress of DB series specimens under failure modes are shown in Figure 13, and hysteresis curves and skeleton curves are shown in Figures 14 and 15, respectively.

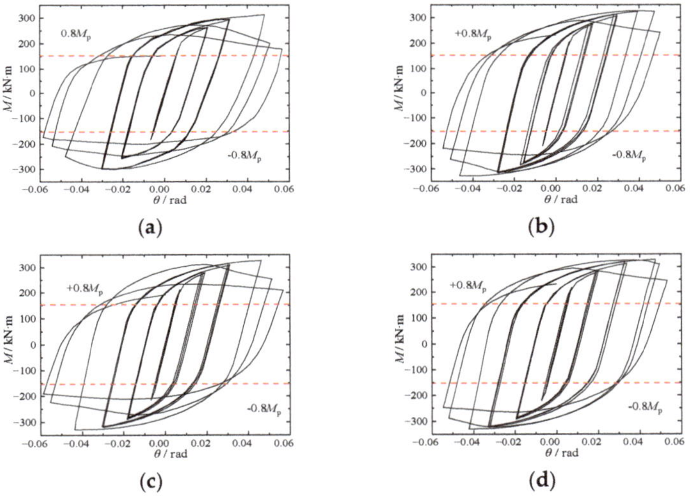

Figure 14. Hysteresis curves of DB series specimens: (**a**) DB-1; (**b**) DB-2; (**c**) DB-3; (**d**) DB-4.

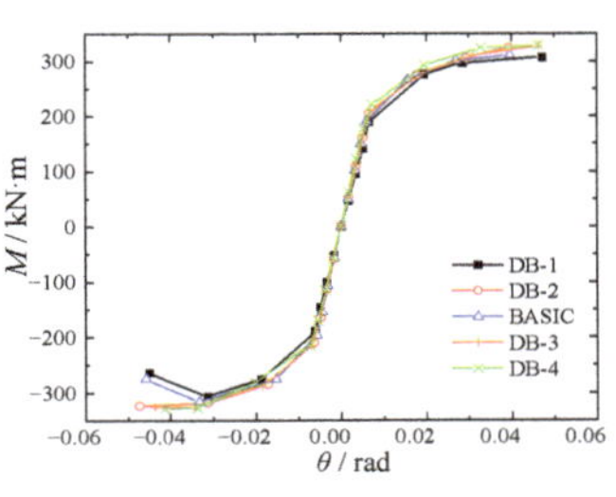

Figure 15. Skeleton curves of DB specimens.

For the DB-1 specimen, larger stress was concentrated at the welding of the beam flange and the end-plate because the thinner end-plate exceeded the maximum stress that the end-plate could suffer. The DB-1 specimen was destroyed earlier than the other three specimens. The plastic strain was mainly concentrated on the U-shaped connector and the column flange at the end of loading. The failure modes of DB-2 to DB-4 specimens were similar to the BASIC specimen, and significant buckling deformation occurred on the beam flange and web at the end of the stiffener. The hysteresis curves of the DB-2 to DB-4 specimens were relatively full, and the plastic deformation and energy dissipation capacity were better.

When the end-plate thickness was varied from 18 to 22 mm, 18 to 24 mm, 18 to 26 mm, and 18 to 30 mm, the maximum moment was increased by 4.0%, 4.2%, 8.3%, and 8.7%, respectively, and the ductility coefficient of specimens was reduced by 1.9%, 2.2%, 8.4%, and 14.7%, respectively. It could be seen that the ductility of the beam end was significantly affected by the end-plate thickness.

Under the same load, the relative rotation of the beam and column would be inhibited by the thicker end-plate, and tensile failure would occur at the bolt. The plastic hinge would arise in the joint because of the thinner end-plate and the joint would fail before the beam. The calculated results of the end-plate thickness could satisfy the seismic demands.

5.3. The Axial Compression Ratio of the Column

The axial compression ratio varied with values of 0.3, 0.4, 0.6 and 0.8. ZY-1, ZY-2, ZY-3 and ZY-4 specimens were established, respectively. The contours of the Von Mises stress of ZY series specimens under failure modes are shown in Figure 16, and hysteresis curves and skeleton curves are shown in Figures 17 and 18, respectively.

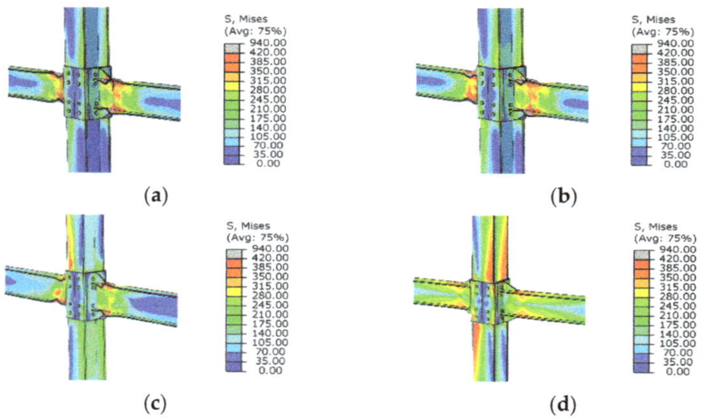

Figure 16. The contour of the Von Mises stress of ZY specimens under failure modes: (**a**) ZY-1; (**b**) ZY-2; (**c**) ZY-3; (**d**) ZY-4.

When the axial compression ratio increased from 0.3 to 0.8, the failure modes of ZY series specimens vastly differed with growth in the axial compression. When the axial compression ratio was less than 0.6, the failure mode of the specimen was consistent with the BASIC specimen.

The high-stress region and the severe buckling deformation appeared at the beam far from the panel zone. No obvious deformation occurred in the panel zone, and the 'strong joint' characteristic was shown in these specimens. When the axial compression ratio was greater than 0.6, the high-stress region was shifted from the beam flange to the column flange, and the failure mode of these joints was changed from beam failure to joint and column failure.

The maximum moment was reduced by 1.1%, 1.4%, 2.1%, and 2.5%, respectively, and the ductility coefficient of specimens was decreased by 0.4%, 4.4%, 4.4%, and 6.5%, respectively, when the axial compression ratio was varied from 0.25 to 0.3, 0.25 to 0.4, 0.25 to 0.6 and 0.25 to 0.8. It could be seen that the axial compression ratio of the column is negatively correlated with the ductility. However, it had little effect on the specimen's yielding moment and maximum moment.

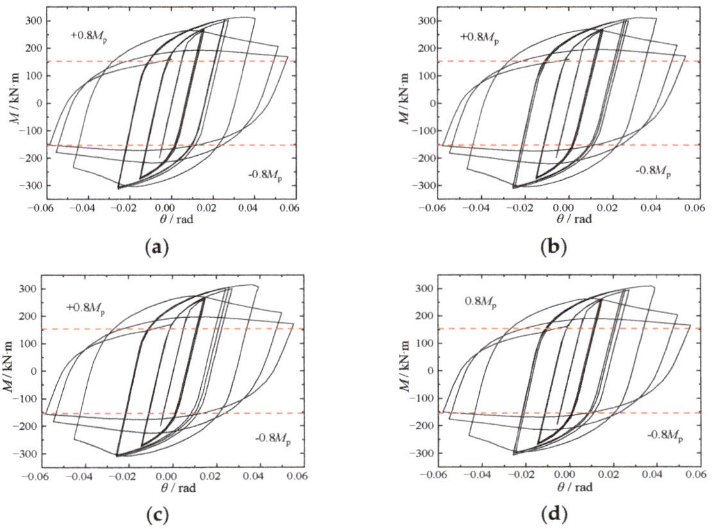

Figure 17. Hysteresis curves of ZY series specimens: (**a**) ZY-1; (**b**) ZY-2; (c) ZY-3; (**d**) ZY-4.

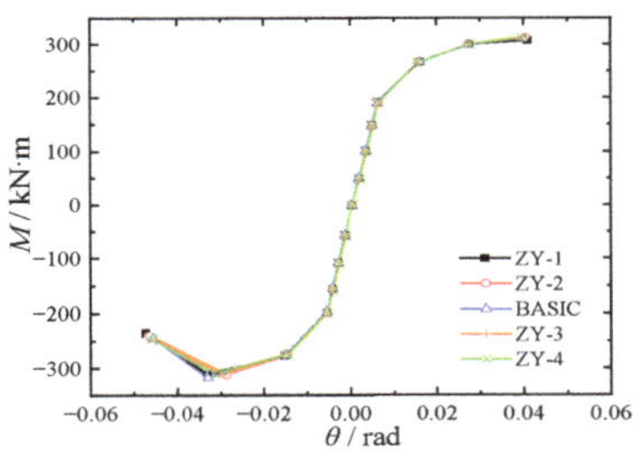

Figure 18. Skeleton curves of ZY specimens.

The seismic behavior of the joint would be reduced with the increase in the axial compression ratio. When the axial compression ratio of the column of the specimen exceeded 0.6, the local buckling deformation of the column flange occurred before the beam flange. Therefore, the axial compression ratio should not exceed 0.6.

5.4. The Linear Stiffness Ratio of the Beam to Column

The beam section size was changed to HN300 × 150 × 6.5 × 9, HN400 × 200 × 8 × 13, HN450 × 200 × 9 × 14, and HN500 × 200 × 10 × 16. BLS-1, BLS-2, BLS-3, and BLS-4 specimens were established, respectively. The linear stiffness ratio of the beam to the column varied with values of 0.32, 1.09, 1.45, and 2.10. The contours of the Von Mises stress of BLS series specimens under failure modes are shown in Figure 19, and the hysteresis and skeleton curves are shown in Figures 20 and 21, respectively.

When the linear stiffness ratio of the beam to column increased from 0.32 to 2.10, failure modes with significant differences appeared on the BLS series specimens. The failure mode was consistent with the BASIC specimen when the linear stiffness ratio of the beam to the column was less than 1.5. When the linear stiffness ratio of the beam to the column was greater than 1.5, the high-stress region was transferred from the beam flange to the vicinity of the column flange in the panel zone. In addition, the failure was concentrated on the column and the U-shaped connector web. The stress value reached the maximum stress value, and the plastic deformation was more significant in these regions.

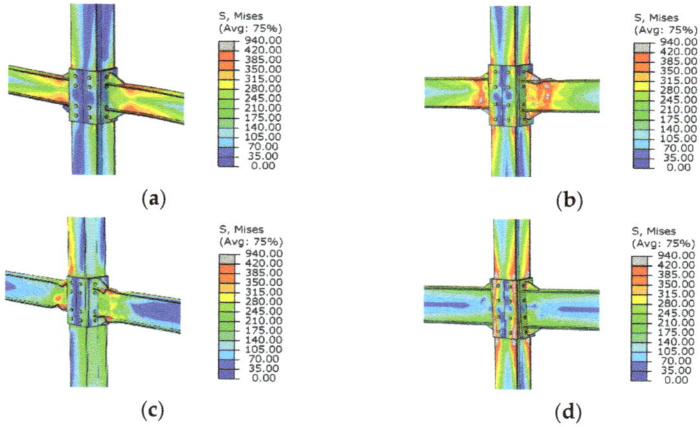

Figure 19. The contour of the Von Mises stress of BLS specimens under failure modes: (**a**) BLS-1; (**b**) BLS-2; (**c**) BLS-3; (**d**) BLS-4.

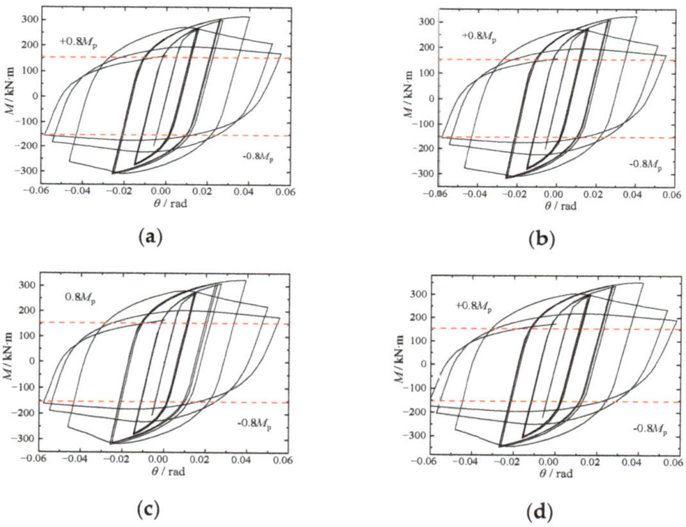

Figure 20. Hysteresis curves of BLS series specimens: (**a**) BLS-1; (**b**) BLS-2; (**c**) BLS-3; (**d**) BLS-4.

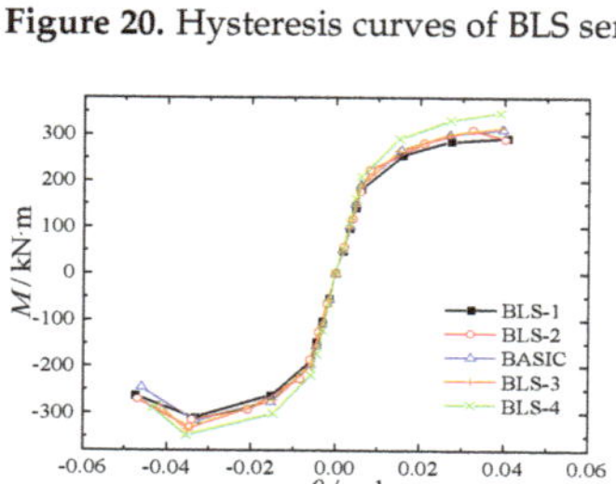

Figure 21. Skeleton curves of BLS specimens.

The maximum moment was increased by 6.3%, 8.0%, and 11.9%, respectively, and the ductility coefficient of specimens was reduced by 4.4%, 14.9%, and 21.1%, respectively, when the linear stiffness ratio of the beam to the column was varied from 0.32 to 1.05,

0.32 to 1.49 and 0.32 to 2.10. It could be seen that the maximum moment of the beam was significantly affected by the linear stiffness ratio of the beam to column, and the linear stiffness ratio of the beam to column was negatively correlated with the ductility behavior.

The failure mode of the specimen was ideal when the linear stiffness ratio of the beam to the column was less than 1.5. Considering the inevitable defects in practical engineering, the linear stiffness ratio of the beam to the column was limited to 1.5.

5.5. Classification of Beam-to-Column Joints

According to the definition of stiffness classification in EC3 [1], the types of joints of each specimen are classified by the calculation of the skeleton curves, as shown in Figure 22.

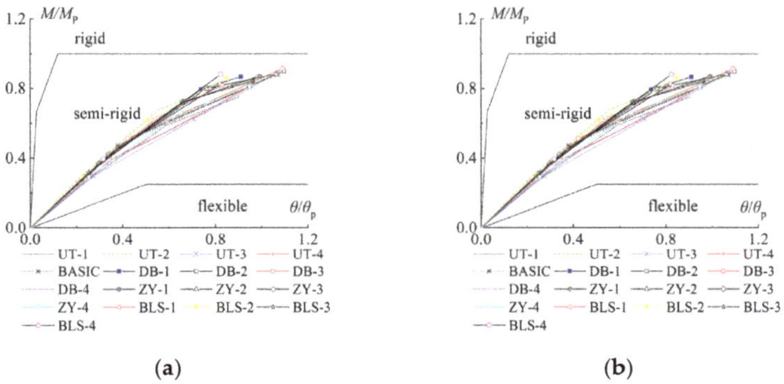

Figure 22. Joint classification: (**a**) non-sway frame; (**b**) sway frame.

The value of the maximum plastic rotation θ_p is calculated by Equation (5).

$$\theta_p = M_p l_b / (E_s I_b), \tag{5}$$

where θ_p represents the value of the maximum plastic rotation; l_b represents the span of the beam; E_s represents elastic modulus; I_b represents the moment of inertia of the beam section.

As shown in Figure 22, each joint is a typical semi-rigid connection based on the stiffness classification, whether in the frame with or without lateral displacement. According to the classification of connection strength, each joint's M/MP value is between 0.25 and 1.0, which means that the type of each joint is an under-strong connection [30], and sufficient rotation capacity is available on each joint. The calculated results demonstrated that the novel kind of joints exhibit better stiffness, ductility, and bearing capacity when compared to rigid and hinged joints, thereby showing superior seismic behavior.

6. Conclusions

Based on the study of the seismic behavior of the BASIC specimen by FEM analysis, the effects of parameters are further analyzed in detail. Based on the results, the following conclusions have been drawn.

(1) The end-plate connection of the steel beam to the weak axis of an H-shaped column with a U-shaped connector is a typical semi-rigid joint with excellent rotation and deformation capacity. Setting the stiffener in the panel zone can keep the plastic hinge far away from the panel zone and avoid the stress concentration of the beam flange and the end-plate welding. The failure mode of the BASIC specimen satisfies the requirements of 'strong joint and weak member'.

(2) The seismic behavior of the joint is affected by the thickness of the U-shaped connector. Plastic deformation occurs in the panel zone if the U-shaped connector is too thin. Conversely, if the U-shaped connector is overly thick, it results in a reduction in joint ductility. The appropriate end-plate thickness can not only make full use of the

bearing capacity of bolts and the end-plate but also ensure that ductility in the joint is superior. The ductility of the joint decreases with the increase in the axial compression ratio. When the linear stiffness ratio of the beam to the column is less than 1.5, plastic deformation is formed at the beam flange and web away from the panel zone, which is an ideal failure mode.

(3) The parameter values are proposed as follows: the thickness of the U-shaped connector t_u should be in the range of $1.15t_f$ to $1.30t_f$; the thickness of the end-plate of the joint can be calculated according to the corresponding specification [28]; the axial compression ratio should not exceed 0.6; the linear stiffness ratio of the beam to the column is limited to 1.5.

Author Contributions: Conceptualization, S.N. and S.W.; methodology, S.N.; software, S.N., S.W. and S.C.; validation, S.N., S.C., W.W., X.J., Y.L. and M.L.; formal analysis, S.N.; investigation, S.N.; resources, S.N.; data curation, S.N.; writing—original draft preparation, S.N. and S.C.; writing—review and editing, S.N., S.C. and W.W.; visualization, S.N.; supervision, S.N.; project administration, S.N.; funding acquisition, S.N. All authors have read and agreed to the published version of the manuscript.

Funding: This paper is supported by the Key Research and Development Program of Shaanxi (No.2021SF-519, No.2024GX-YBXM-374).

Data Availability Statement: The data presented in this study are available on request from the corresponding author. The data are not publicly available due to patent protection.

Conflicts of Interest: The authors declare no conflicts of interest.

Nomenclature

Δ	The horizontal displacement
H	The height from the loading point to the bottom of the column
R_1	The reaction force of the end of the left beam
l	The distance from the beam end to the surface of the U-shaped connector
N	Preliminary constant axial load
M_p	The plastic moment of the whole section of the beam
θ_p	The value of the maximum plastic rotation
M_y	The yielding moment of the joint
M_{max}	The maximum moment of the joint
M_u	The ultimate moment of the joint
θ_y	The inter-story drift under the yielding moment of the joint
θ_{max}	The inter-story drift under the maximum moment of the joint
θ_u	The inter-story drift under the ultimate moment of the joint
μ	The rotational ductility coefficient (=θ_u/θ_y).
R_{ki}	The initial stiffness
N_t	The design value of the tensile force of a high-strength bolt
e_w	The distance from the center of the bolt to the web
e_f	The distance from the center of the bolt to the flange plate
f	The design strength of the end-plate
l_b	The span of the beam
E_s	Elastic modulus
I_b	The moment of inertia of the beam section

References

1. European Committee for Standardization. *Eurocode 3: Design of Steel Structural. Part 1-8: Design of Joints;* Wiley: London, UK, 2005.
2. Wang, J.; Chen, L. Experimental investigation of extended end-plate joints to concrete-filled steel tubular columns. *Constr. Steel Res.* **2012**, *79*, 56–70. [CrossRef]
3. Wang, J.; Zhang, L.; Spencer, B.F. Seismic response of extended end-plate joints to concrete-filled steel tubular columns. *Eng. Struct.* **2013**, *49*, 876–892. [CrossRef]
4. Shi, G.; Shi, Y.; Li, S.; Wang, Y. Cyclic tests of semi-rigid end-plate connection in multi-story steel frames. *Build. Struct.* **2005**, *26*, 74.

5. Shi, Y.J.; Wang, M.; Wang, Y.Q. Analysis on the behavior of steel frame end-plate connections. *Eng. Mech.* **2011**, *28*, 51–58.
6. Zhao, B.D. Extended End-Plate Bolted Connections Behavior Study of Steel Frame Structure. Ph.D. Thesis, Xi'an University of Architecture and Technology, Xi'an, China, 2006.
7. Ghobarah, A.; Osman, A.; Korol, R.M. Behavior of extended end-plate connections under cyclic loading. *Eng. Struct.* **1990**, *12*, 15–27. [CrossRef]
8. Sumner, E.A.; Murray, T.M. Behavior of extended end-plate moment connections subject to cyclic loading. *Struct. Eng.* **2002**, *128*, 501–508. [CrossRef]
9. Cabrero, J.M.; Bayo, E. The semi-rigid behavior of three-dimensional steel beam-to-column joints subjected to proportional loading. part I: Experimental evaluation. *Constr. Steel Res.* **2007**, *63*, 1241–1253. [CrossRef]
10. Nie, S.F.; Wang, S.; Liu, B. Study on the seismic behavior of weak-axis end-plate joint of reinforced beam to column with U-shaped connector. *Prog. Steel Build. Struct.* **2023**, *25*, 38–48.
11. Lu, L.F.; Liu, Y.; Li, K.S. Finite element analysis on the seismic behavior of DRBS weak-axis connection with box-shaped region. *Prog. Steel Build. Struct.* **2019**, *21*, 87–96.
12. Lu, L.F.; Xu, Y.L.; Zheng, H. Investigation of composite action on seismic performance of weak-axis column bending connections. *Constr. Steel Res.* **2017**, *129*, 286–300. [CrossRef]
13. Wang, L.L. Study on Seismic Behavior and Design Method of Widen Beam-End Flange Weak Axis Connection with Box-Shaped Panel Zone of I-Section Column. Master's Thesis, Chang'an University, Xi'an, China, 2018.
14. Tagawa, H.; Gurel, S. Application of steel channels as stiffeners in bolted moment connections. *Constr. Steel Res.* **2005**, *61*, 1650–1671. [CrossRef]
15. Kukreti, A.P.; Shou, F.-F. Eight-bolt endplate connection and its influence on frame behavior. *Eng. Struct.* **2006**, *28*, 1487–1493. [CrossRef]
16. Kim, S.; Kim, S.; Ju, K.Y. Strength evaluation of beam–column connection in the weak axis of H-shaped column. *Eng. Struct.* **2007**, *30*, 1699–1710. [CrossRef]
17. Shim, H.J.; Lee, E.T.; Kim, S.B.; Kim, S.S. Development and Performance Evaluation of Weak-Axis Column Bending Connections for Advanced Constructability. *Int. J. Steel Struct.* **2014**, *14*, 369–380. [CrossRef]
18. Nassiraei, H.; Rezadoost, P. Initial stiffness, ultimate capacity and failure mechanisms of tubular X-joints with external ring or plates at fire conditions. *Ships Offshore Struct.* **2024**, *236*, 1–23. [CrossRef]
19. Rahgozar, N.; Rahgozar, N. Experimental and Numerical Investigation on Flexural Strengthening of Precast Concrete Corbel Connections with Fiber-Reinforced Plastic Sheet. *Buildings* **2024**, *14*, 387. [CrossRef]
20. Nie, S.F.; Wu, Y.F.; Wang, J.Q. Study on seismic properties of weak-axis joint of beam-column with web opening strengthened by casing. *Earthq. Resist. Eng. Retrofit.* **2019**, *41*, 22–30.
21. Nie, S.F.; Ye, M.N.; Wu, Y.F. Seismic behavior of square tube confined steel-reinforced concrete column-RC ring beam joint. *Archit. Civ. Eng.* **2019**, *36*, 84–91.
22. *GB 50017-2017*; Standard for Design of Steel Structures. China Architecture & Building Press: Beijing, China, 2017.
23. Xu, Y.L.; Su, Y.X.; Shang, Y.F. Finite element analysis of rib reinforced weak-axis connector in steel frame. *Eng. Mech.* **2021**, *38*, 31–38.
24. Li, S.F. Bolted end-plate connection of steel structure. *J. Build. Struct.* **1998**, *8*, 24–26.
25. *JGJ/T 101-2015*; Specification for Seismic Test of Building. China Architecture & Building Press: Beijing, China, 2015.
26. Lou, G.B.; Li, G.Q.; Lei, Q. Recent developments in the behavior and design of high-strength bolted end-plate connections(I). *Prog. Steel Build. Struct.* **2006**, *8*, 8–21.
27. Lou, G.B.; Li, G.Q.; Lei, Q. Recent developments in the behavior and design of high-strength bolted end-plate connections(II). *Prog. Steel Build. Struct.* **2006**, *8*, 16–23.
28. *ANSI/AISC 341-10*; Seismic Provisons for structural steel Buidings. American Institute of Steel Construction: Chicago, IL, USA, 2010.
29. *GB 51022-2015*; Technical Code for Steel Structure of Light-weight Building with Gabled Frames. China Architecture & Building Press: Beijing, China, 2015.
30. *GB 50011-2016*; Code for Seismic Design of Buildings. China Architecture & Building Press: Beijing, China, 2016.

buildings

MDPI

Article

Finite Element Investigation of a Novel Cold-Formed Steel Shear Wall

Zhiqiang Xie [1], Ye Bi [1], Ying Fan [1,*], Chengwei Gao [1], Xiangdong Zhang [2], Yin Feng [2], Daxing Zhou [2] and Lei Dong [2]

[1] School of Civil and Transportation Engineering, Beijing University of Civil Engineering and Architecture, Beijing 102616, China; xiezhiqiang@bucea.edu.cn (Z.X.); 2108590022032@stu.bucea.edu.cn (Y.B.); 2108590022088@stu.bucea.edu.cn (C.G.)

[2] China Railway Construction Group Co., Beijing 100070, China; zhangxiangdong@ztjs.cn (X.Z.); fengyin.ztjs@crcc.cn (Y.F.); zhoudaxing@ztjs.cn (D.Z.); donglei@ztjs.cn (L.D.)

* Correspondence: 2108590022034@stu.bucea.edu.cn

Abstract: In this paper, a novel corrugated steel sheet central sheathed cold-formed steel (CCS-CFS) shear wall is proposed. This shear wall can address the problems of low shear strength and ductility in conventional cold-formed steel (CFS) shear walls caused by screw connection failure and eccentric sheet arrangement. A numerical simulation method for the novel shear wall was developed and verified through cyclic loading test results of two full-size shear wall specimens. Parameter analysis was then conducted to investigate the effects of screw spacing, sheet thickness ratio, and aspect ratio on the seismic performance of these shear walls, accompanied by design recommendations. The results indicated that this innovative shear wall configuration can effectively resolve the connection failure between the frame and the sheet. Furthermore, the CCS-CFS shear wall can effectively improve shear strength, energy dissipation capacity, and ductility. The developed numerical simulation method can accurately capture the hysteretic properties and failure modes of shear walls. In addition, it can address the shortcomings in conventional models that neglect the mixed hardening characteristic of steel and metal damage criteria, resulting in inaccurate simulation results and unrealistic buckling modes. The principal failure modes observed in the novel shear wall were identified as the plastic buckling of corrugated steel sheathing and the distortional buckling of the end stud. Reducing the screw spacing has a limited impact on its shear strength. It is recommended that the sheet thickness ratio of the CCS-CFS shear wall should be greater than 2.0, while the aspect ratio can be relaxed to 10:4.

Keywords: CCS-CFS shear wall; seismic performance; numerical simulation method; parametric analysis; design recommendations

1. Introduction

Earthquakes have a significant impact on the structural safety of buildings [1,2]. Many scholars have conducted research in this field. Deng, E et al. [3] presented an experimental–numerical investigation on an innovative fully prefabricated liftable connection (FPLC) for modular steel buildings (MSBs). The results indicate that the FPLC had a satisfactory deformation capacity and sufficient ductility under earthquake conditions. Yang, L et al. [4] proposed a displacement-amplified mild steel bar joint damper to solve some dampers that cannot give full play to the energy dissipation effect during small earthquakes. Wei, J et al. [5] carried out research on the seismic performance of concrete-filled steel tubular (CFST) composite columns with ultra-high performance concrete (UHPC) plates. These structures can withstand multiple earthquakes and have demonstrated good seismic performance.

A cold-formed steel (CFS) structure is a type of green building structure that has been widely used in low-rise residential [6,7] and office buildings due to its characteristics of light weight, high strength, industrialization and a high degree of assembly. In recent years, propelled by advancements in the construction industry, the CFS structure system has become a promising prospect for development in mid and high-rise constructions.

The main lateral stability and strength in CFS structures are provided by CFS shear walls. Therefore, the seismic performance of a shear wall [8–10] critically influences the seismic performance of the whole structure. Research by Niari et al. [11] and Feng et al. [12] revealed that the screw connection between the sheet and the frame in a single-sided CFS shear wall was prone to tilt and slip in the process of stress. This resulted in a noticeable pinching effect of the hysteretic curves and poor energy dissipation of components. In addition, screw connection failure was a brittle form of failure which meant that the seismic potential of components was not fully exploited, resulting in low shear strength and ductility of the shear wall. To address the above problems, a self-piercing rivet (SPR) connection and a hybrid connection (SPR and screws) were developed by Xie et al. [13,14]. Through experimental investigations, the researchers proposed methods for calculating the shear and tensile strength for these two types of joints, significantly improving the connection performance of shear walls. Therefore, improving the screw connection performance or introducing new connection types is an effective way to enhance the seismic performance of CFS shear walls.

An important reason for the low shear strength and ductility of CFS shear walls is the eccentric force caused by the single-sided placement of sheathing. Research conducted by Rizk and DaBreo et al. [15,16] revealed that the off-center positioning of the sheet subjected the studs of the wall to torsional moments. Furthermore, shear buckling deformations of the sheathing led to the sheet pulling over the heads of the screws, resulting in a loss of the out-of-plane constraint of the sheet on the steel frame. This decreased the compression strength and anti-collapse capabilities of the shear wall. Given these findings, experimental investigations on double-sheathed shear walls were completed by Santos [17]. The results indicated that while this new configuration effectively improved the torsional strength and stiffness of the wall, the challenge of low ductility due to premature screw connection failures remained. Subsequently, Wang et al., Briere et al. and Rogers et al. [18–20] innovatively proposed the center-sheathed CFS shear wall. Although this innovation partially solved the aforementioned problems, shear bearing capacity and stiffness were not substantially enhanced, owing to the low lateral stiffness of steel plates and the susceptibility to out-of-plane buckling.

For this reason, Yu et al. and Zhang et al. [21,22] developed a CFS shear wall sheathed with corrugated steel sheets. They found that corrugated steel sheathing demonstrated superior shear resistance compared to conventional steel sheathing. However, the main failure modes were screw connection failure and end stud buckling, with limited improvement in ductility.

Developing a numerical model of a CFS shear wall is an important way to simulate and predict its failure mode and hysteretic properties. Ngo et al. [11] proposed a numerical simulation method for CFS shear walls. However, the rigidity assumption of the sheathing was adopted in the modeling process, which meant the relative displacement of the sheet was ignored. As a result, the peak load and initial stiffness of the simulation results were too small, and the decreasing section of the hysteretic curve was not obvious. Niari [23] and Xu et al. [24] conducted finite element analysis of CFS shear walls by considering geometric and material nonlinearity, but this did not effectively solve the aforementioned problems. Additionally, Xu assumed the steel sheet to be an ideal elastic–plastic material, but this led to plastic deformation occurring easily in the middle of the steel sheet or the screw connection area at the corner of the wall. This significantly reduced the computational efficiency of the model.

To sum up, while the existing numerical model of a CFS shear wall considered material and geometric non-linearity to enhance its accuracy, the utilization of an ideal elastic–plastic constitutive model for steel proved inadequate. This manifested in challenges such as overly high initial stiffness and the absence of a clearly delineated decline section within the simulated hysteretic curve. Therefore, the numerical simulation of a CFS shear wall that considers the characteristics of steel mixed strengthening and metal damage criteria requires further development.

According to the aforementioned scholars' research on traditional CFS shear walls, it is evident that the conventional cold-formed steel (CFS) shear wall exhibited weak resistance and low ductility caused by screw connection failure and eccentric sheet arrangement. To achieve a stronger and more ductile CFS shear wall, an innovative corrugated steel sheet central sheathed cold-formed steel (CCS-CFS) shear wall was conceived. Based on this, a refined numerical simulation method of the CCS-CFS shear wall was developed by considering the characteristics of steel mixed strengthening and metal damage criteria. The three issues of undersized peak load, excessive initial stiffness, and the decreasing section of the hysteretic curve were not obvious and could be effectively resolved. The effectiveness of the method was verified by cyclic loading test results of two full-size shear wall specimens. Following the validated finite element model, a comprehensive finite element analysis was conducted on parameters such as screw spacing, sheet thickness ratio, and aspect ratio. The impact of these parameters on the seismic performance and failure mechanism of the shear wall was systematically studied.

2. Design and Experimental Program of CCS-CFS Shear Wall Configuration

2.1. Innovative Configuration of the CCS-CFS Shear Wall

As a response to the limited strength and ductility of conventional CFS shear walls, this paper presents an innovative CCS-CFS shear wall with corrugated steel sheet central sheathing, as depicted in Figure 1. In this configuration, two back-to back corrugated steel sheets were centrally confined within the framing, providing a substantial increase in both shear resistance and ductility in comparison with conventional shear walls. The corrugated steel sheets were sandwiched by built-up vertical studs along the edges and horizontal members located at the base and top of the wall. Additionally, L-shaped connectors were used to connect the studs and the horizontal members, enhancing the overall stability of the shear wall through improved component connections. Furthermore, a symmetrical arrangement of hold-downs on both sides of the wall was employed to enhance connection performance and prevent hold-down failure or damage.

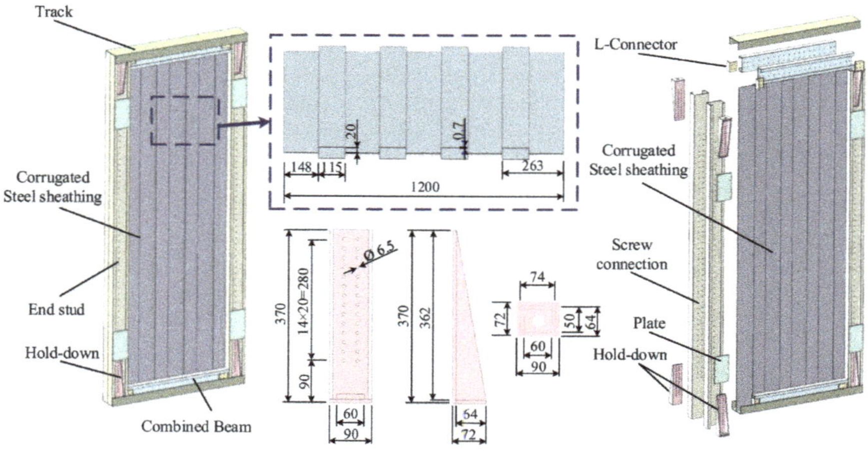

Figure 1. Schematic diagram of CCS-CFS shear wall (unit: mm).

Compared with the conventional shear wall, the CCS-CFS shear wall offers the following advantages: (1) The two back-to-back corrugated steel sheets are centrally confined within the framing and sandwiched by built-up vertical studs along the edges and horizontal members located at the base and top of the wall. While ensuring a tight fit between the sheet and frame, the shear surface of the screws can be increased to avoid premature failure or pulling out of the sheets. This enhancement can significantly improve the shear strength and stiffness of the wall. (2) Placing the sheet in the middle of the frame serves a dual purpose. Firstly, by eliminating the torsion effect of the column caused by the eccentricity of the panel, it enhances the wall's compression and collapse resistance. Secondly, it generates extra connection space and augments the number of connections between the sheet and frame. This reinforcement amplifies the skin effect of the sheet, enlarges the slip space of the connection, and ultimately enhances the ductility and seismic energy dissipation capabilities of the wall. (3) As a high-performance lateral force resistance component, the novel shear wall can be applied not only in multi-story steel structures but also in low-story CFS structures.

2.2. Specimen Design

Two full-size CCS-CFS shear wall specimens were designed and manufactured. The height and width of all specimens were 3000 mm and 1200 mm, respectively. The transverse brace comprised double C-shaped steel with 140 mm web × 55 mm flange × 25 mm lip × 2 mm thickness, fastened together by two lines of screws at intervals of 50 mm along the length of the web. The end studs consisted of two coupled C-shaped steel with 140 mm web × 75 mm flange × 25 mm lip × 2 mm thickness, fastened together by two lines of screws at intervals of 50 mm along the length of the web. The track comprised double U-shaped steel with 160 mm web × 70 mm flange × 2 mm thickness attached by two lines of SPRs at intervals of 50 mm along the length of the web. The interval between the two lines of screws was 40 mm. The cross-section size of the corrugated steel sheet was CS 1200 mm × 148 mm × 115 mm × 20 mm, with a thickness of 0.7 mm. The end studs of specimen SW-1 were not equipped with plates, while the end studs of specimen SW-2 were equipped with plates. Figure 2 presents the detailed configurations of the specimens.

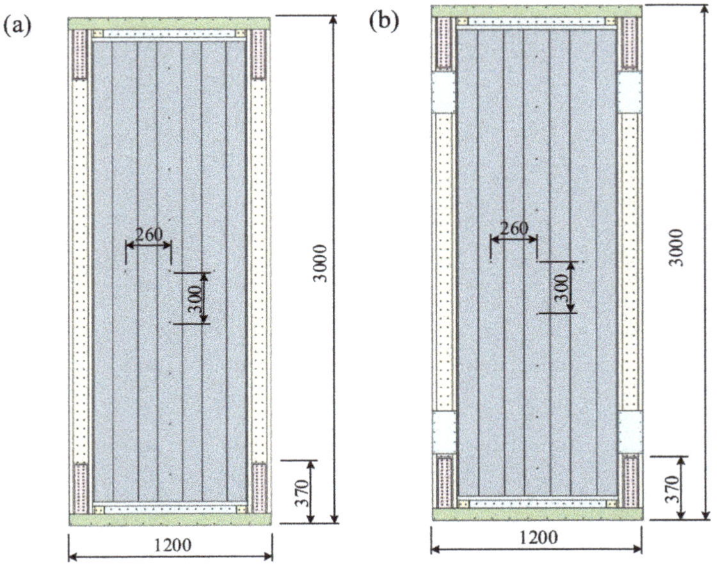

Figure 2. Configurations of specimens: (**a**) specimen SW-1; (**b**) specimen SW-2 (unit: mm).

ST5.5-grade screws were used for the shear wall. Due to the different thickness of the connected sheet, the length of the screw between the combined transverse brace and the corrugated steel sheet was 25 mm; the length between the transverse brace and the wave trough was 70 mm; and the length between the end studs and the corrugated steel sheet was 32 mm. ST6.3-grade screws were employed between the hold-downs and the steel frame. The failure modes, hysteretic curves, skeleton curves, and test results of two specimens were obtained by conducting cycle loading tests.

2.3. Material Properties

In accordance with Chinese Standard GB/T 228.1 [25], three coupons were tested for each component the results of which are presented in Table 1. The strength-to-yield ratio (f_u/f_y) of coupons was greater than 1.2 and the elongation did not fall below 10%. The ductility of materials satisfied the provision of North American specification AISI S100-16 [26].

Table 1. Coupon test results for CFS components.

Component	Steel Strength Grade	Steel Thickness t (mm)	Yield Strength f_y (MPa)	Tensile Strength f_u (MPa)	f_u/f_y	Elongation S (%)
Corrugated steel	Q235	0.7	238	320	1.34	33.67
Steel frame	Q355	2.0	330	433	1.31	23.72

2.4. Test Results Analysis

2.4.1. Failure Modes

At low load levels, both specimens exhibited elastic shear buckling of the corrugated steel sheet. When horizontal displacement reached 8mm, plastic buckling of the corrugated steel sheet was observed. At this point, the shear strength of the two specimens decreased, and the decline of specimen SW-1 was larger. As the horizontal displacement continued to increase, the shear strength of the two specimens gradually increased and a tension field appeared along the diagonal of specimens, as depicted in Figures 3a and 4a. Simultaneously, the two flanges at the bottom of the left end stud buckled and the specimens entered the failure stage. When the wall shear strength dropped to 85% of the peak load, the test was terminated and the specimens were considered destroyed, as illustrated in Figures 3 and 4. To summarize, the main failure modes of the novel shear wall were the plastic buckling of corrugated steel sheathing and the distortional buckling of the end stud. Most notably, the distortional buckling of the end stud was the key factor affecting the shear strength of the shear wall.

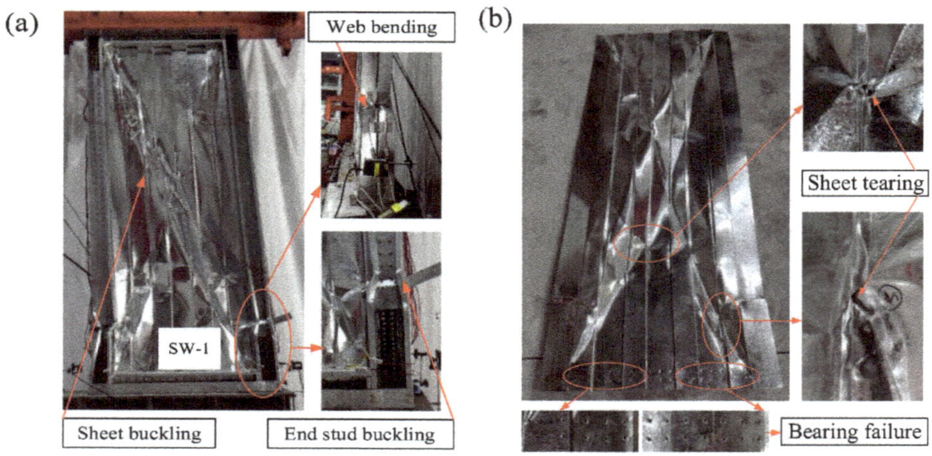

Figure 3. Failure modes of SW-1: (**a**) specimen SW-1; (**b**) corrugated steel sheet.

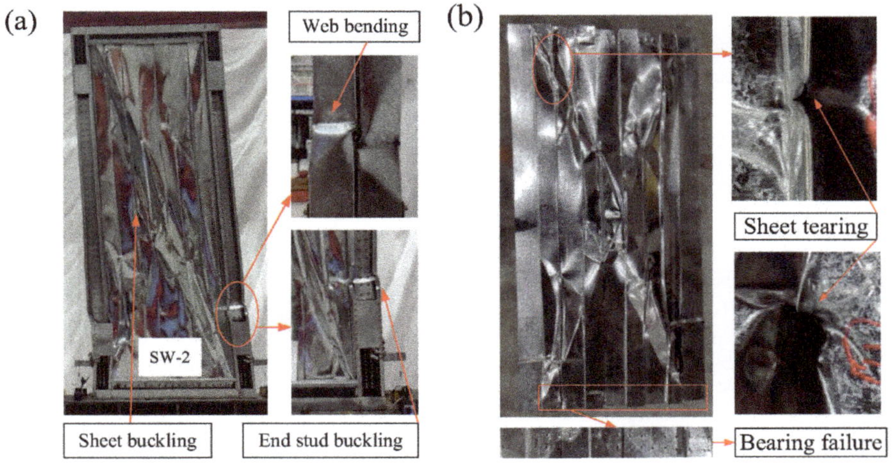

Figure 4. Failure modes of SW-2: (**a**) specimen SW-2; (**b**) corrugated steel sheet.

2.4.2. Hysteresis Curves

Figure 5 presents the hysteresis curves for each specimen. The test results are presented in Table 2. As depicted in Figure 5, the hysteresis curves of specimens SW-1 and SW-2 were similar. When the peak load was achieved, the shear strength of the specimens significantly degraded. The plastic hinge formed by the buckling of the end stud and the plastic deformation and tearing of the steel sheet increased the slip of the hysteresis curve, making the "pinched" phenomenon more obvious. Compared with specimen SW-1, the deformation capacity and cumulative energy dissipation capacity of specimen SW-2 increased by 16.4% and 16.6%, respectively. This was because the addition of plates to the end studs significantly enhanced its local flexural and compressive strength, effectively delaying the buckling and failure process. However, the shear strength and stiffness of the wall primarily depended on the sheet and steel frame, while the plate served merely to locally strengthen the end stud without directly enhancing the shear strength of the sheet or steel frame. Thus, the impact of adding plates on the shear strength and stiffness of the shear wall was relatively small.

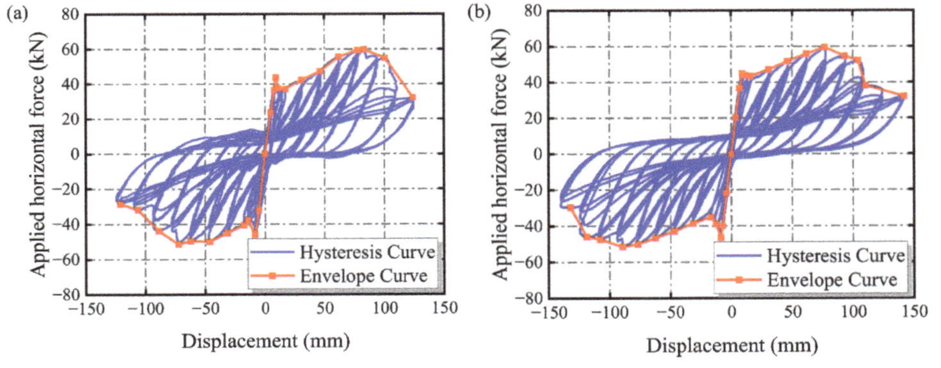

Figure 5. Hysteretic curves of the specimens: (**a**) specimen SW-1; (**b**) specimen SW-2.

Table 2. Test results.

Specimen Label	Amplitude	K_e (kN/mm)	Δ_y (mm)	P_y (kN)	Δ_{max} (mm)	P_{max} (kN)	Δ_u (mm)	μ (--)	E (J)	E_c (--)
SW-1	Positive	4.60	40.35	45.28	83.09	59.73	104.83	2.60	2.37	0.69
	Negative	6.28	20.48	41.54	72.19	51.44	88.89	4.34	2.09	0.57
	Average	5.44	30.42	43.41	77.64	55.59	96.86	3.18	2.23	0.63
SW-2	Positive	5.20	28.81	47.07	77.01	59.40	105.20	3.65	2.59	0.65
	Negative	5.52	32.10	39.47	89.50	51.62	120.46	3.75	2.61	0.79
	Average	5.36	30.46	43.27	83.26	55.51	112.83	3.70	2.60	0.72

2.4.3. Comparison with the Conventional CFS Shear Wall

Based on the test results obtained by Xie et al. [27], this section presents a comparison of the hysteretic curves and seismic performance indexes between the conventional CFS shear wall (Figure 6a) and specimen SW-2 (Figure 6b). The results of the comparison of hysteresis curves are displayed in Figure 6c. As depicted in Figure 6, specimen SW-2 demonstrated an approximately 60% increase in steel usage compared with the conventional CFS shear wall. Nevertheless, notable enhancements were observed in its shear stiffness, shear strength, and cumulative energy consumption, with increases of 267%, 208%, and 175%, respectively. In summary, the seismic performance of the CCS-CFS shear wall markedly improved compared with the conventional CFS shear wall. Hence, it is advisable to consider employing the CCS-CFS shear wall as a prospective lateral force resistance solution within a multi-layer CFS structure system.

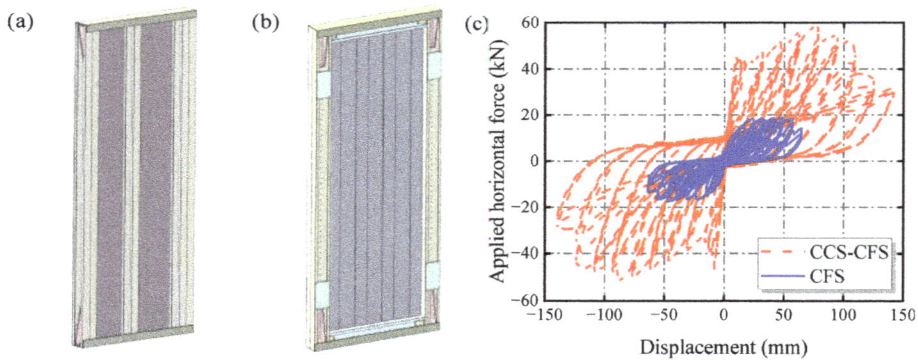

Figure 6. Comparison between the CFS shear wall and the CCS-CFS shear wall: (**a**) CFS shear wall; (**b**) specimen SW-2; (**c**) hysteretic curves.

3. Numerical Methodology

3.1. Modeling of Shear Walls

3.1.1. Element Choice and Mesh Size

As depicted in Figure 7, the S4R shell element was adopted to model the steel frame and sheet sheathing in ABAQUS [28]. The dimensions of the model were consistent with the specimen. During finite element modeling, an odd number of section points is specified throughout the shell thickness when integrating properties during analysis. Due to the relatively uniform thickness of materials within the wall, ABAQUS employs five section points throughout the thickness of a homogeneous shell. A study by Schafer et al. [29] revealed that the local, distortional and global buckling failure modes can be simulated more accurately with medium or suitable mesh. Therefore, the mesh size for the steel frame was set at 20 mm, while the mesh size for the sheet sheathing was 30 mm (Figure 7b).

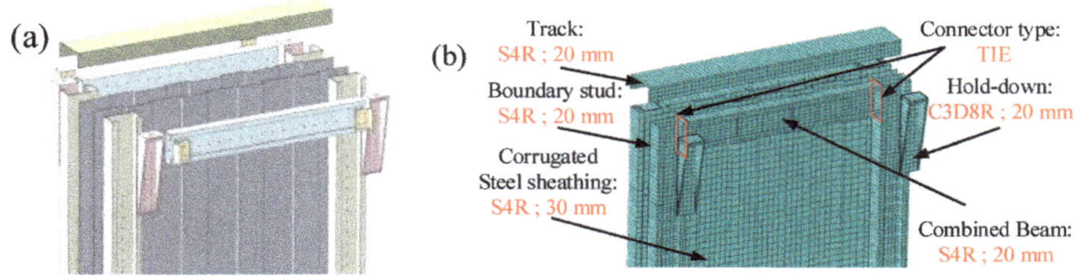

Figure 7. Finite element model of the CCS-CFS shear wall: (**a**) schematic diagram; (**b**) finite element model.

3.1.2. Material Modeling

According to the von Mises yield criteria [30], the steel exhibited the characteristics of mixed strengthening, as displayed in Figure 8a. To accurately simulate the hysteresis curve of the actual steel material, the combined module in ABAQUS and cycle hardening material properties were employed to simulate the steel under cycle loads. Material parameters were fitted using reference [31], and the results are detailed in Table 3.

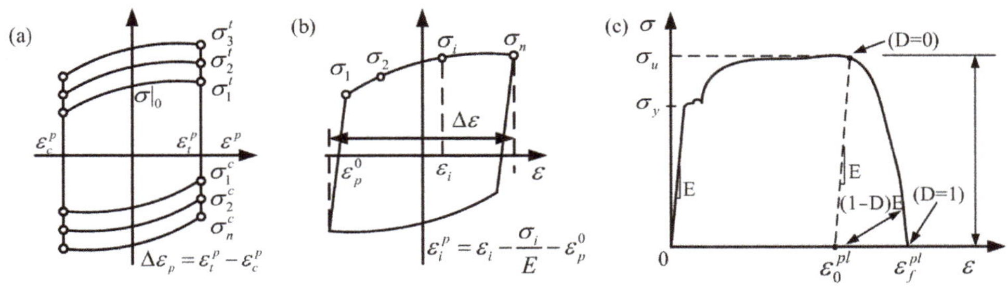

Figure 8. Material model of steel: (**a**) isotropic strengthening model of steel; (**b**) steel dynamic strengthening model; (**c**) true stress–strain curve of steel tensile.

Table 3. Material parameter calibration.

Steel Strength Grade	$\sigma\|_0$ (N/mm^2)	Q_∞ (N/mm^2)	b_{iso}	$C_{kin,1}$ (N/mm^2)	γ_1	$C_{kin,2}$ (N/mm^2)	γ_2	$C_{kin,3}$ (N/mm^2)	γ_3	$C_{kin,4}$ (N/mm^2)	γ_4
Q235	238	21	1.2	6050	175	5050	120	3050	25	1000	35
Q355	330	21	1.2	8000	175	6800	120	2850	35	1450	30

Note: $\sigma|_0$ represents the stress at zero equivalent plastic strain; Q_∞ is the maximum change value of yield surface. b_{iso} is the ratio that the yield surface changes with increasing plastic strain. $C_{kin,k}$ and γ_k are constants, proofread according to test results.

Furthermore, steel material properties were determined according to coupon tests. The nominal stress (σ_{nom}) and strain (ε_{nom}) data were obtained from the material test (Table 1), and transformed into the input true stress (σ_{ture}) and strain (ε_{ture}) based on Equations (1)–(3) [32], as displayed in Figure 8b. Figure 8c depicts the true stress–strain curve of steel under tension. In order to achieve a more precise simulation of the failure behavior of the shear wall, ductile damage properties were incorporated into the material constitutive model. Additionally, the damage evolution sub-term of the steel damage evolution path was included.

$$\varepsilon_{nom} = \frac{l - l_0}{l_0} = \frac{l}{l_0} - 1 \quad (1)$$

$$\sigma_{ture} = \sigma_{nom}(1 + \varepsilon_{nom}) \tag{2}$$

$$\varepsilon_{ture} = \ln(1 + \varepsilon_{nom}) \tag{3}$$

3.1.3. Geometric Nonlinearity

Geometric nonlinearity in structural analysis pertains to the requirement of generating a new equilibrium equation based on the deformed state of a structure following significant deformation. Failure to account for this phenomenon can lead to substantial calculation errors due to the evolving shape and behavior of the structure under load. For example, consider a cantilever beam loaded vertically at the tip. If the tip deflection is small, the analysis can be considered as being approximately linear. However, if the tip deflections are large, the shape of the structure, and hence its stiffness, changes. Therefore, the geometric nonlinearity of the cantilever beam needs to be considered [33].

Consequently, because each member of the CCS-CFS shear wall was a shell element, the presence of geometric nonlinearities became particularly pronounced. When subjected to bending loads, these shell elements were prone to significant deformations, resulting in substantial displacements and rotations at critical buckling points within the structure. Therefore, the influence of geometric nonlinearity on shear wall must be considered in finite element analysis.

3.1.4. Modeling of the Screw Connection

The screw connection in the shear wall can be classified into three categories based on the varying base steel thicknesses: (1) connection between the track and the transverse brace (2 + 2 mm); (2) connection between the corrugated steel sheet and the end stud (2 + 0.7 + 0.7 + 2 mm); (3) connection between the corrugated steel sheet and the transverse brace (2 + 0.7 mm). Based on the test results, it was found that the screw connection between the track and the transverse brace remained essentially intact, with no instances of connection failure. Consequently, it was inferred that the joint exhibited rigidity, and was consequently modeled using a TIE constraint in the simulation. Slight hole wall expansion was evident in the connection between the corrugated steel sheet and the steel frame, indicating that a relative slip occurred between the two components. Thus, it was assumed that the joint was hinged and was simulated by a Cartesian connector (Figure 9). Because the combined plate thickness of the corrugated steel sheet and end stud differed from that of the corrugated steel sheet at the wave crest and transverse brace, two different constitutive models of screws were required for simulation. The constitutive model and calculation parameters are presented in Figure 10 and Table 4.

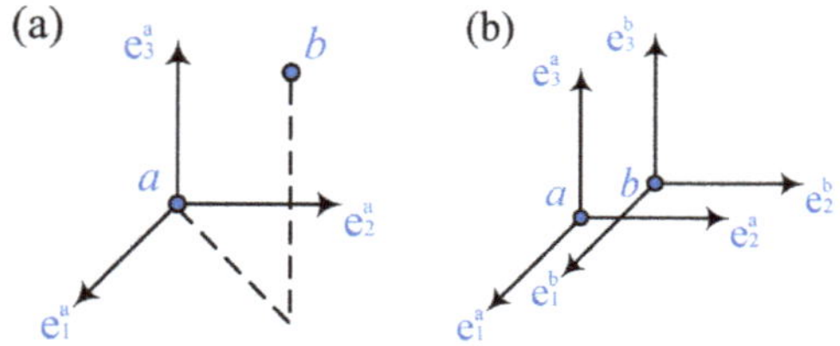

Figure 9. Translational and rotational relations of the Cartesian connector: (**a**) translational connection; (**b**) rotary connection.

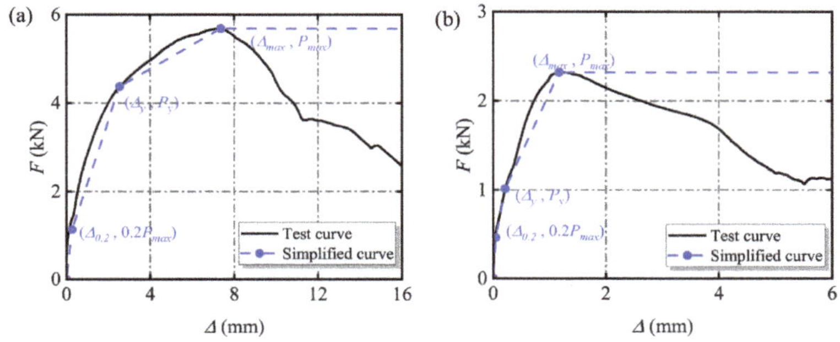

Figure 10. Tested properties of the screw connection: (**a**) constitutive model I; (**b**) constitutive model II.

Table 4. Tested properties of the screw connection.

Connection Object	Constitutive Model	$\Delta_{0.2}$ (mm)	$0.2P_{max}$ (kN)	K_e (kN/mm)	Δ_y (mm)	P_y (kN)	Δ_{max} (mm)	P_{max} (kN)
End stud + Corrugated steel	I	0.511	1.136	2.226	2.553	4.376	7.371	5.682
Beam + Corrugated steel	II	0.047	0.464	9.910	0.214	1.013	1.168	2.319

Figure 11 presents a simplified diagram of the hold-downs in the finite element model. During the cyclic loading tests, no damage was observed on the hold-downs, and they consistently remained within the elastic range. Therefore, the solid unit C3D8R was employed to simulate the hold-downs, and the constraint between it and the end stud was simplified to the TIE constraint.

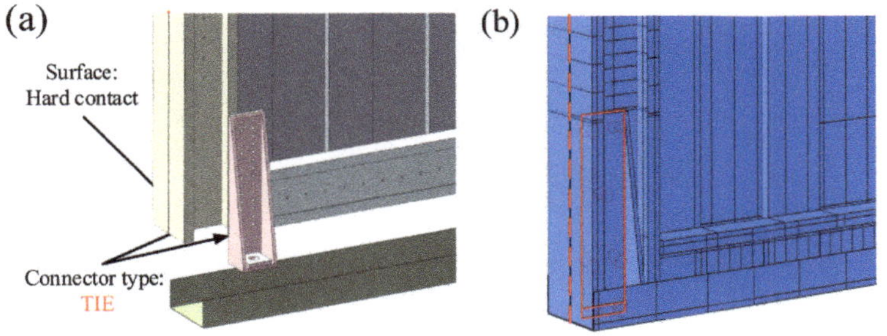

Figure 11. Simplified diagram of the hold-downs in finite element model: (**a**) drawing of the hold-downs; (**b**) finite element simulation of the hold-downs.

3.1.5. Boundary Conditions and Loading Mode

According to the test, the simulation of the boundary conditions of the CCS-CFS shear wall was primarily divided into the following: (1) The relationship between the top track and the loading beam. The web plate of the upper track was coupled to the reference point RP-1, and the translation degree of freedom in the loading direction was constrained. (2) Preventing out-of-plane instability of the wall during loading. The translation degree of freedom in the external direction of the wall was restricted. (3) The connection between the bottom track and the beam. The translation degree of freedom between the bottom track and the hold-downs was restricted. (4) Surface-to-surface contact with "hard-contact" behavior in a normal direction was introduced to simulate the interfaces between the frame members and the sheet. Additionally, the tangent behavior of interfaces was defined as frictionless, as depicted in Figure 12. To improve convergence in the analysis and acquire a hysteresis curve featuring a descending branch, we employed a monotonic loading scenario utilizing displacement-control loading within the model.

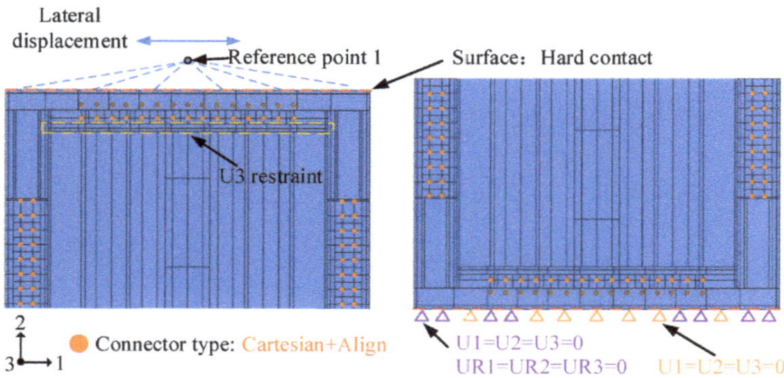

Figure 12. Cross-sectional dimensions of steel members.

3.2. Model Validation

3.2.1. Comparison of Failure Mode

As depicted in Figure 13, the stress concentration in both specimens was primarily located in the compressed side studs (near the bottom hold-downs), which was consistent with the test results. In addition, the tension bands of both corrugated steel sheets were formed along the diagonal of the wall, which was in agreement with the test results. In summary, the failure mode observed in the finite element model aligned with the experimental results, thus validating the method used for modeling the CCS-CFS shear wall.

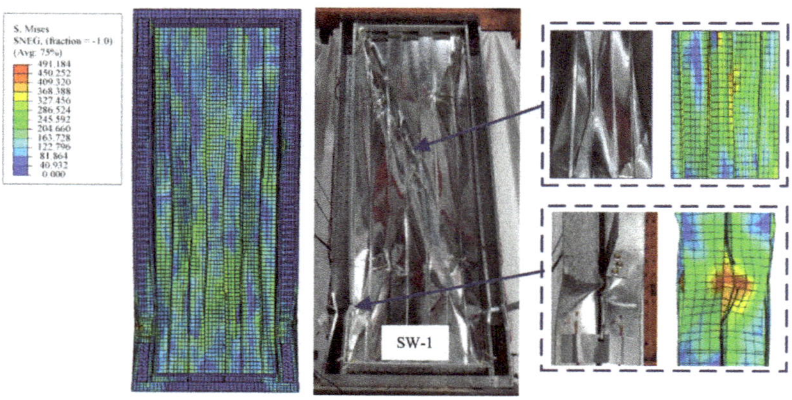

Figure 13. Comparison of failure modes between the finite element and test for specimen SW-1.

3.2.2. Comparison of Load–Displacement Curve

The test and finite element analysis hysteretic curves are displayed in Figure 14. For specimen SW-1, the hysteretic curve derived from the finite element analysis closely matched the test results, exhibiting a distinct decline section. Compared with the test results, the displacement of the finite element analysis curve in the falling section was larger. This variation could be attributed to the TIE constraint applied in certain connections of the wall, potentially resulting in slightly increased ductility of the shear wall in the simulation compared with the test specimen. However, as can be seen in Table 5, it is evident that the relative error between the test results and the finite element results was maintained within 12%, which meets the requisite standards in the field of civil engineering and enables subsequent parameter analysis.

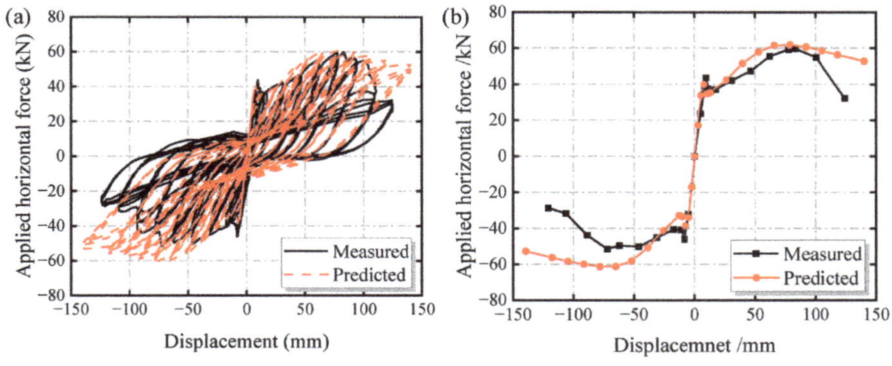

Figure 14. Comparison of test and finite element load–displacement curves: (**a**) hysteretic curves; (**b**) backbone curves.

Table 5. Comparison of eigenvalues between the finite element analysis and test results.

Specimen Label		K_e (kN/mm)	Δ_y (mm)	P_y (kN)	Δ_{max} (mm)	P_{max} (kN)
SW-B-1	FE analysis	5.81	33.96	47.79	78.60	61.46
	Test	5.44	30.42	43.41	77.64	56.59
	Error (%)	6.80	11.63	10.09	1.24	8.61

4. Parametric Analyses

4.1. Influence of Screw Spacing on the Seismic Performance of the CCS-CFS Shear Wall

To investigate the effect of the screw spacing on the seismic performance of the CCS-CFS shear wall, five finite element models were designed with different screw spacings 50 mm, 75 mm, 100 mm, 125 mm and 150 mm [34,35]. The skeleton curves for these models are presented in Figure 15.

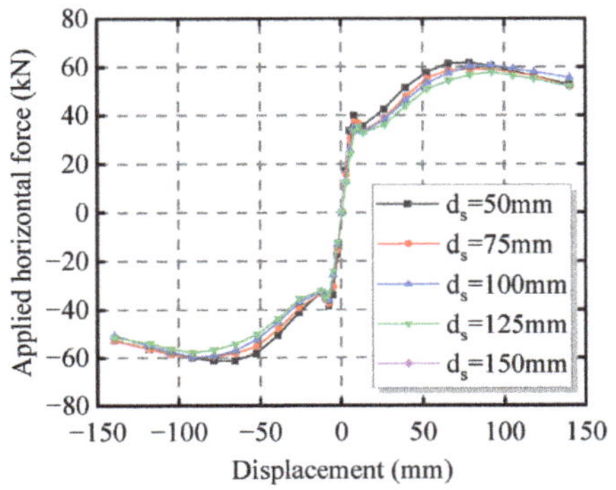

Figure 15. Skeleton curves with various screw spacings.

As illustrated in Figure 15, the impact of screw spacing on the seismic performance of the shear wall was relatively constrained. The finite element analysis results are listed in Table 6. As detailed in Table 6, when screw spacing increased in the range of 50 mm to 150 mm, the shear strength of the shear wall decreased linearly with a change range of −7.5%. By contrast, the ductility increased linearly in line with the increase in screw spacing, with a change range of approximately 39.7%. This is attributable to the innovative steel frame of the CCS-CFS shear wall, which modified the failure mode observed in conventional CFS shear walls. Consequently, the shear strength of the shear wall primarily relied on the steel frame and the sheet. However, as the screw spacing increased, the number of screws in the tension band of the sheet decreased. This may reduce the connection strength between the sheet and steel frame, while also decreasing the resistance of the screw group, resulting in heightened deformability. Therefore, it is highly recommended that the screw spacing be set at 100 mm to ensure the seismic performance of the shear wall and facilitate construction.

Table 6. Finite element calculation results under different screw spacings.

Specimen Label	Screw Spacing (mm)	Yield Displacement (mm)	Yield Load (kN)	Maximum Displacement (mm)	Maximum Load (kN)	Stiffness (kN/mm)	Ductility
SW-1-F1	50	33.96	47.79	78.60	61.46	5.81	3.50
SW-1-F2	75	35.76	47.35	78.74	60.19	5.40	4.13
SW-1-F3	100	42.11	46.76	91.87	59.74	4.95	4.21
SW-1-F4	125	42.39	45.60	91.90	57.96	4.71	4.71
SW-1-F5	150	42.50	45.54	91.92	57.11	4.12	4.89

4.2. Influence of Sheet Thickness Ratio on the Seismic Performance of the CCS-CFS Shear Wall

According to the North American specification AISI S100-16 [26], the frame thickness (t_s) not be less than the sheet thickness (t_p). Referring to the methods for calculating the shear strength of self-tapping screws proposed in this specification, two cases of sheet thickness ratio $t_s/t_p \geq 2.5$ and $1.0 \leq t_s/t_p < 2.5$ were analyzed. The corresponding parameters and specimen labels are listed in Table 7.

Table 7. Parameters and numbers of finite element models with different member thicknesses.

Sheet Thickness Ratio (t_s/t_p)		Specimen Label	Height (m) × Width (m)	Stud (mm)	Track (mm)	Sheet (mm)	Remark
$t_s/t_p \geq 2.5$	4.00	SW-1-F6	3.0 × 1.2	2.0	2.0	0.5	The frame thickness is fixed, the sheet thickness is variable.
	3.33	SW-1-F7	3.0 × 1.2	2.0	2.0	0.6	
	2.86	SW-1-F1	3.0 × 1.2	2.0	2.0	0.7	
	2.50	SW-1-F8	3.0 × 1.2	2.0	2.0	0.8	
	2.86	SW-1-F1	3.0 × 1.2	2.0	2.0	0.7	The sheet thickness is fixed, the frame thickness is variable.
	3.57	SW-1-F9	3.0 × 1.2	2.5	2.5	0.7	
	4.29	SW-1-F10	3.0 × 1.2	3.0	3.0	0.7	
$1 \leq t_s/t_p < 2.5$	2.22	SW-1-F11	3.0 × 1.2	2.0	2.0	0.9	The frame thickness is fixed, the sheet thickness is variable.
	2.00	SW-1-F12	3.0 × 1.2	2.0	2.0	1.0	
	1.43	SW-1-F13	3.0 × 1.2	2.0	2.0	1.4	
	1.33	SW-1-F14	3.0 × 1.2	2.0	2.0	1.5	
	1.25	SW-1-F15	3.0 × 1.2	2.0	2.0	1.6	
	1.00	SW-1-F16	3.0 × 1.2	2.0	2.0	2.0	
	1.00	SW-1-F17	3.0 × 1.2	0.7	0.7	0.7	The sheet thickness is fixed, the frame thickness is variable.
	1.14	SW-1-F18	3.0 × 1.2	0.8	0.8	0.7	
	1.29	SW-1-F19	3.0 × 1.2	0.9	0.9	0.7	
	1.43	SW-1-F20	3.0 × 1.2	1.0	1.0	0.7	
	1.71	SW-1-F21	3.0 × 1.2	1.2	1.2	0.7	
	2.14	SW-1-F22	3.0 × 1.2	1.5	1.5	0.7	

Figures 16 and 17 depict the skeleton curves of models with sheet thickness ratios $t_s/t_p \geq 2.5$ and $1 \leq t_s/t_p < 2.5$, respectively. As shown in Figure 17, when the sheet thickness ratio $t_s/t_p \leq 2.0$, the shear wall was prone to brittle failure, which should be avoided in practical engineering applications. Therefore, it is suggested that in the structural design, the sheet thickness ratio t_s/t_p should be greater than 2.0 to ensure the seismic performance of the CCS-CFS shear wall. The simulation results of $t_s/t_p > 2.0$ are listed in Table 8. Figure 18 presents the skeleton curves of the models with sheet thickness ratio $t_s/t_p > 2.0$.

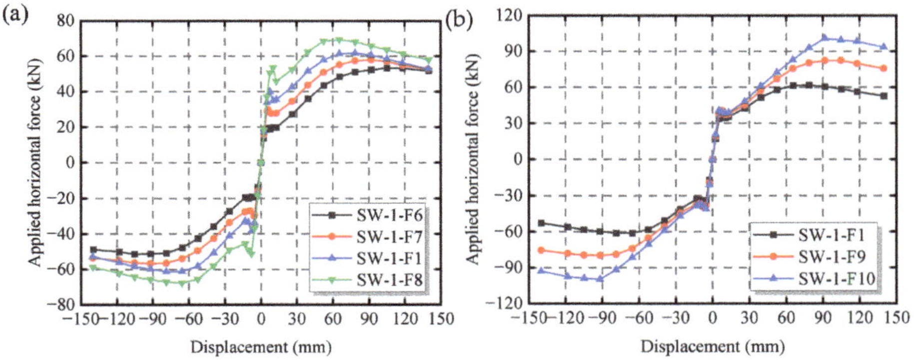

Figure 16. Skeleton curves with $t_s/t_p \geq 2.5$: (**a**) the specimen with various sheet thicknesses; (**b**) the specimen with various frame thicknesses.

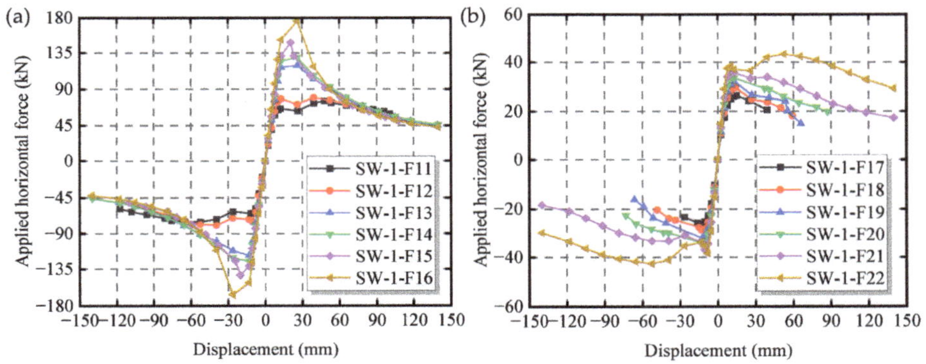

Figure 17. Skeleton curves with $1 \leq t_s/t_p < 2.5$: (**a**) the specimen with various sheet thicknesses; (**b**) the specimen with various frame thicknesses.

Table 8. Finite element calculation results under different member thicknesses.

Sheet Thickness Ratio (t_s/t_p)	Specimen Label	Δ_y (mm)	P_y (kN)	Δ_{max} (mm)	P_{max} (kN)	K_e (kN/m)	μ
$t_s/t_p > 2.0$	SW-1-F6	56.78	44.97	104.98	53.31	5.27	4.00
	SW-1-F7	44.70	46.40	91.87	57.81	5.40	4.03
	SW-1-F1	33.96	47.79	78.60	61.73	5.81	4.17
	SW-1-F8	26.23	52.41	65.62	69.26	5.93	4.99
	SW-1-F11	19.01	63.42	52.46	75.35	7.59	5.40
	SW-1-F22	16.24	36.81	52.50	43.23	5.69	6.10
	SW-1-F1	33.96	47.79	78.60	61.73	5.81	4.03
	SW-1-F9	51.18	65.77	84.82	82.26	6.40	3.50
	SW-1-F10	64.09	81.82	91.88	100.82	7.62	3.17

As detailed in Table 8, when the sheet thickness ratio $t_s/t_p > 2.0$ and the frame thickness was fixed and the sheet thickness was increased, the shear strength, shear stiffness, and ductility coefficient of the shear wall correlated positively with the sheet thickness. Conversely, both the yield displacement and maximum displacement correlated negatively with the sheet thickness. The increases in shear capacity, shear stiffness, and ductility coefficient were 41.34%, 44.02% and 35.00%, respectively. Thus, when $t_s/t_p > 2.0$, the stiffness and ductility of the shear wall can be effectively improved by increasing the sheet thickness.

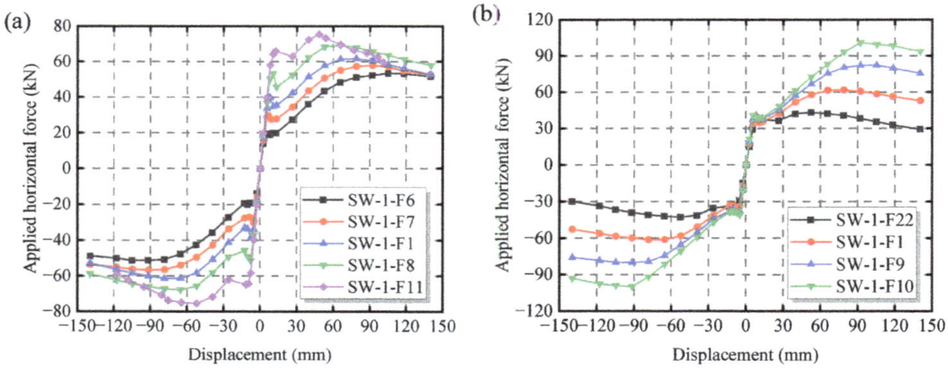

Figure 18. Skeleton curves with $t_s/t_p > 2.0$: (**a**) the specimen with various sheet thicknesses; (**b**) the specimen with various frame thicknesses.

However, when the sheet thickness was fixed, the shear strength and maximum displacement of the shear wall correlated positively with the frame thickness. In contrast, the shear stiffness and ductility coefficient exhibited minimal variation. The increases in maximum displacement and shear strength were 75.01% and 133.22%, respectively. Thus, when $t_s/t_p > 2.0$, increasing the frame thickness can effectively improve the shear strength and deformation capacity of the shear wall.

In summary, the seismic performance of the CCS-CFS shear wall was significantly influenced by the sheet thickness ratio, with an optimal ratio existing between the frame and the sheet. When $t_s/t_p \leq 2.0$, the shear wall was prone to brittle failure. Therefore, it is suggested that the sheet thickness ratio of the CCS-CFS shear wall should exceed 2.0. When $t_s/t_p > 2.0$ and under the same parameters, increasing the frame thickness exerted a more significant impact on the shear strength of the shear wall than increasing the sheet thickness. By contrast, increasing the sheet thickness had a more pronounced effect on enhancing the shear stiffness and ductility of the shear wall compared to increasing the frame thickness.

4.3. Influence of Aspect Ratio on the Seismic Performance of the CCS-CFS Shear Wall

The North American specification AISI S400 [36] stipulates that the aspect ratio of a shear wall should not exceed 2. When the ratio is greater than 2:1 and less than 4:1, the shear strength should be reduced by a reduction factor of 2 w/h. However, whether this reduction method that considers the influence of aspect ratio is applicable to the CCS-CFS shear wall is not particularly clear. Therefore, seven finite element analysis models with aspect ratios of 1.08, 1.33, 1.74, 2.50, 3.00, 3.96 and 4.45 were established to study the influence of aspect ratio on the seismic performance of the CCS-CFS shear wall.

Failure modes of CCS-CFS shear walls with different aspect ratios are displayed in Figure 19. As indicated in Figure 19, the stress concentration in both specimens was primarily located in the compressed side studs (near the bottom hold-downs) and the tension band of the sheet. Therefore, the failure mode of shear wall specimens was primarily manifested as distortion buckling of the end stud on the compressive side and buckling of the sheet. When the aspect ratio of the wall increased, the distortion degree of the compression side stud was weakened, and the number of shear deformation half-waves of the sheet was reduced.

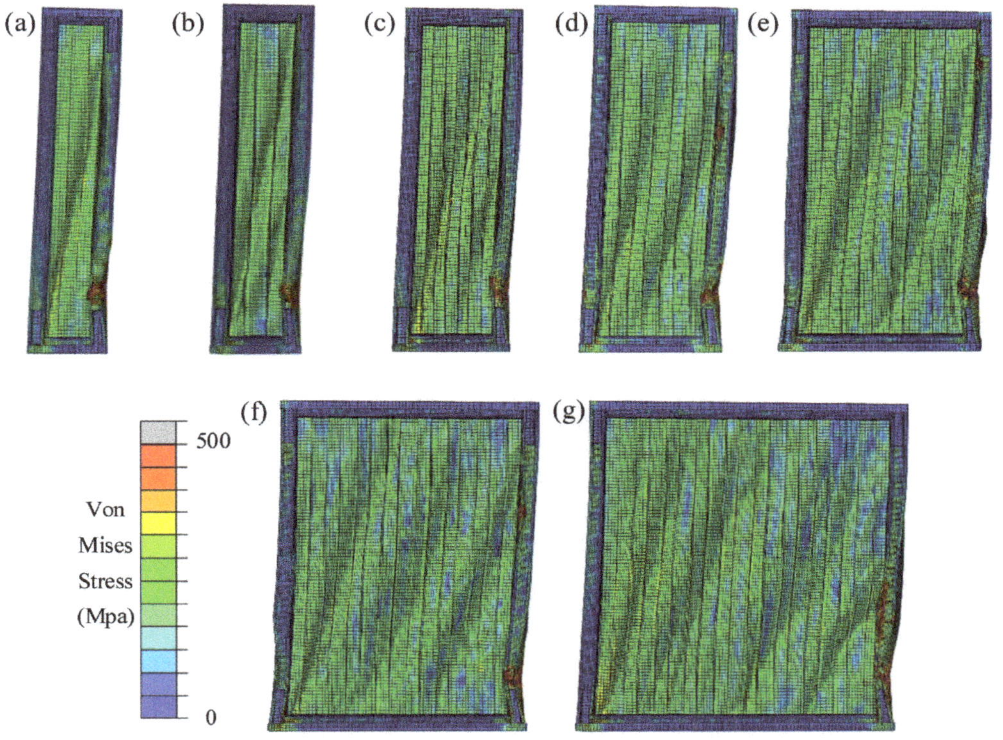

Figure 19. Failure modes of CCS-CFS shear walls under different aspect ratios: (**a**) 4.45:1; (**b**) 3.96:1; (**c**) 3.00:1; (**d**) 2.5:1; (**e**) 1.74:1; (**f**) 1.33:1; (**g**) 1.08:1.

Figure 20 depicts the skeleton curves of CCS-CFS shear walls with different aspect ratios. The test and finite element analysis results of shear walls with different aspect ratios are listed in Table 9. When the ratio increased from 1.08 to 4.45, the maximum displacement and shear strength of shear walls changed by −78.84% and 54.17%, respectively.

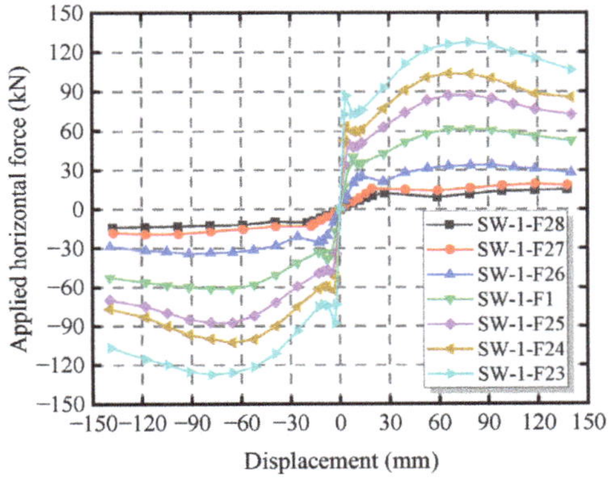

Figure 20. Skeleton curves with different aspect ratios.

Table 9. Parameters and numbers of the finite element model under different aspect ratios.

Specimen Label	Aspect Ratio	Δ_{max} (mm)	P_{max} (kN)	Shear Strength (kN/m)	μ
SW-1-F23	1.08	78.21	137.53	49.51	4.13
SW-1-F24	1.33	78.75	113.06	50.47	4.10
SW-1-F25	1.74	78.73	87.14	50.49	4.14
SW-1-1	2.50	78.60	61.73	51.22	4.17
SW-1-F26	3.00	91.94	34.00	33.66	3.78
SW-1-F27	3.96	118.11	19.34	25.51	2.77
SW-1-F28	4.45	137.56	15.23	22.60	2.35

The relationship between the reduction coefficient of shear strength and the aspect ratio of the CCS-CFS shear wall is illustrated in Figure 21. The reduction factor of shear strength for CCS-CFS shear walls with different aspect ratios is more conservative when determined according to the method recommended by AISI S400. Under reciprocating loading, the shear strength of CCS-CFS shear walls approached unity when the aspect ratio was less than 2.5. To summarize, the effect of aspect ratio on the shear strength of CCS-CFS shear walls can be directly considered using the North American standard AISI S400 [36] and the aspect ratio limit for CCS-CFS shear walls can be relaxed to 10:4.

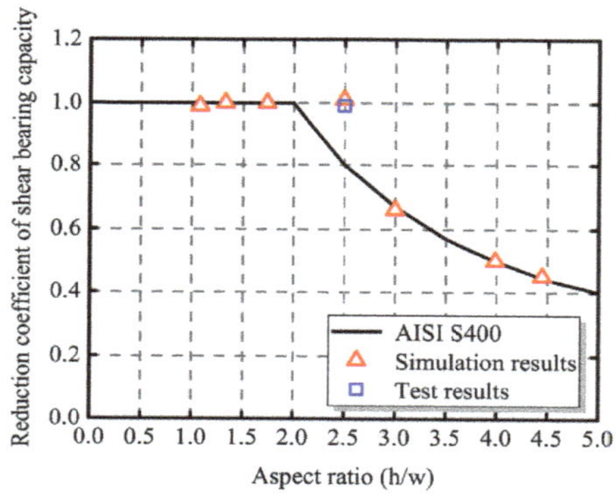

Figure 21. Relationship between reduction coefficient and aspect ratio.

5. Conclusions

An innovative corrugated steel sheet central sheathed cold-formed steel (CCS-CFS) shear wall was proposed in this paper. Based on this, a refined numerical simulation method for the CCS-CFS shear wall was developed and verified through the cyclic loading test results of two full-size shear wall specimens. Furthermore, a comprehensive finite element analysis was conducted on parameters including screw spacing, sheet thickness ratio, and aspect ratio. The main conclusions are as follows.

(1) The CCS-CFS shear wall effectively solved the problem of a connection failure between frame and sheet. Under cyclic load, the main failure modes were the plastic buckling of the corrugated steel sheet and the distortional buckling of the end stud. The method of adding a plate to the side stud considerably improved the deformation ability of the shear wall but had limited influence on its shear strength and stiffness.

(2) Compared with the conventional CFS shear wall, the shear strength, cumulative energy consumption and shear stiffness of the CCS-CFS shear wall were increased by 208%, 175%, and 267%, respectively. Therefore, it is recommended that the CCS-CFS shear wall be employed as a potential lateral force resistance scheme in a multi-layer CFS structure system.

(3) By considering the characteristics of steel mixed strengthening and metal damage criteria, the detailed numerical simulation of the CCS-CFS shear wall developed in this paper can simulate the real failure mode of a shear wall. Furthermore, the finite element analysis results were in good agreement with the test results.
(4) The influence of screw spacing on the seismic performance of the shear wall was relatively limited. Thus, it is highly recommended that the screw spacing be set at 100 mm to ensure the seismic performance of the shear wall and facilitate construction.
(5) To prevent brittle damage to the wall, it is recommended that the sheet thickness ratio of the CCS-CFS shear wall exceed 2.0. Additionally, increasing the frame thickness can effectively enhance the shear strength of the shear wall, while significantly improving the shear stiffness and ductility of the shear wall can be achieved by increasing the sheet thickness.
(6) The aspect ratio exerted a significant influence on both the shear strength and maximum displacement of the CCS-CFS shear wall. The impact on the shear strength of the shear wall can be directly assessed by referencing the North American code AISI S400, but its aspect ratio limit can be relaxed to 10:4.

Author Contributions: Methodology, Z.X. and Y.F. (Ying Fan); Software, Y.B.; Validation, X.Z., Y.F. (Yin Feng), D.Z. and L.D.; Data curation, Z.X., Y.B. and C.G.; Writing—original draft, Y.B. and Y.F. (Ying Fan); Writing—review & editing, Z.X., Y.F. (Ying Fan), C.G., X.Z., Y.F. (Yin Feng), D.Z. and L.D.; Visualization, Z.X.; Supervision, Z.X. All authors have read and agreed to the published version of the manuscript.

Funding: This research is supported by the National Natural Science Foundation of China (Grant No. 52008018), and the Program for Scientific Research of Beijing Municipal Education Commission (Grant No. KM202110016012). The research is also partially supported by the Pyramid Talent Training Project of Beijing University of Civil Engineering and Architecture (Grant No. JDYC20220804), the Cultivation project Funds for Beijing University of Civil Engineering and Architecture (Grant No. X23046), the Project funded by China Railway Construction Group Co., Ltd. (Grant No. LX22-21b) and the BUCEA Post Graduate Innovation Project (Grant No. PG2023030). Any opinions, findings, conclusions, or recommendations expressed in this article are those of the authors and do not necessarily reflect the views of the sponsors.

Data Availability Statement: Data is contained within the article.

Conflicts of Interest: Authors Xiangdong Zhang, Yin Feng, Daxing Zhou and Lei Dong were employed by the company China Railway Construction Group Co. The remaining authors declare that the research was conducted in the absence of any commercial or financial relationships that could be construed as a potential conflict of interest.

References

1. Huang, H.; Yao, Y.; Liang, C.; Ye, Y. Experimental study on the cyclic performance of steel-hollow core partially encased composite spliced frame beam. *Soil Dyn. Earthq. Eng.* **2022**, *163*, 107499. [CrossRef]
2. Zhang, H.; Xiang, X.; Huang, B.; Wu, Z.; Chen, H. Static homotropy response analysis of structure with random variables of arbitrary distributions by minimizing stochastic residual error. *Comput. Struct.* **2023**, *288*, 107153. [CrossRef]
3. Deng, E.; Wang, Y.; Zong, L.; Zhang, Z.; Zhang, J. Seismic behavior of a novel liftable connection for modular steel buildings: Experimental and numerical studies. *Thin-Walled Struct.* **2024**, *197*, 111563. [CrossRef]
4. Yang, L.; Ye, M.; Huang, Y.; Dong, J. Study on Mechanical Properties of Displacement-Amplified Mild Steel Bar Joint Damper. *Iran. J. Sci. Technol. Trans. Civ. Eng.* **2023**. [CrossRef]
5. Wei, J.; Ying, H.; Yang, Y.; Zhang, W.; Yuan, H.; Zhou, J. Seismic performance of concrete-filled steel tubular composite columns with ultra-high performance concrete plates. *Eng. Struct.* **2023**, *278*, 115500. [CrossRef]
6. Hasanali, M.; Roy, K.; Mojtabaei, S.M.; Hajirasouliha, I.; Clifton, G.C.; Lim, J.B.P. A critical review of cold-formed steel seismic resistant systems: Recent developments, challenges and future directions. *Thin-Walled Struct.* **2022**, *180*, 109953. [CrossRef]
7. Hasanali, M.; Mojtabaei, S.M.; Clifton, G.C.; Hajirasouliha, I.; Torabian, S.; Lim, J.B.P. Capacity and design of cold-formed steel warping-restrained beam–column elements. *J. Constr. Steel Res.* **2022**, *190*, 107139. [CrossRef]
8. Deng, R.; Ye, L.; Wang, Y.H.; Li, P.; Shi, Y. Lateral performance of cold-formed steel framed shear walls using slitted sheathing with stiffeners. *Eng. Struct.* **2024**, *302*, 117385. [CrossRef]

9. Wu, J.; Rogers, C. Cold-formed steel centre-sheathed (mid-ply) shear walls of intermediate resistance. *Thin-Walled Struct.* **2023**, *188*, 110834. [CrossRef]
10. Liu, X.; Zhang, W.; Yu, C.; Li, Y.; Jiang, Z.; Yu, S. Experimental study on cold-formed steel shear walls with different corrugated steel sheathings. *J. Constr. Steel Res.* **2022**, *199*, 107639. [CrossRef]
11. Niari, S.E.; Rafezy, B.; Abedi, K. Seismic behavior of steel sheathed cold-formed steel shear wall: Experimental investigation and numerical modeling. *Thin-Walled Struct.* **2015**, *96*, 337–347. [CrossRef]
12. Feng, R.Q.; Zhu, B.; Xu, P.; Qiu, Y. Seismic performance of cold-formed steel framed shear walls with steel sheathing and gypsum board. *Thin-Walled Struct.* **2019**, *143*, 106238. [CrossRef]
13. Xie, Z.; Zhang, W.; Chen, T.; Zhou, D.; Shi, L.; Tang, Y.; Yu, C. Comparative analysis and design method of shear strength for hybrid SPR-SDS joints in thin-walled steel structures. *Structures* **2021**, *33*, 4313–4329. [CrossRef]
14. Xie, Z.; Zhang, A.; Yan, W.; Zhang, Y.; Mu, T.; Yu, C. Study on shear performance and calculation method for self-pierce riveted joints in galvanized steel sheet. *Thin-Walled Struct.* **2021**, *161*, 107490. [CrossRef]
15. Rizk, R. Cold-Formed Steel Frame-Steel Sheathed Shear Walls: Improved Range of Shear Strength Values Accounting for Effect of Full Frame Blocking and Thick Sheathing/Framing Members. Master's Thesis, McGill University, Montreal, QC, Canada, 2017.
16. DaBreo, J.; Balh, N.; Ong-Tone, C.; Rogers, C.A. Steel sheathed cold-formed steel framed shear walls subjected to lateral and gravity loading. *Thin-Walled Struct.* **2014**, *74*, 232–245. [CrossRef]
17. Santos, V. Higher Capacity Cold-Formed Steel Sheathed and Framed Shear Walls for Mid-Rise Buildings. Master's Thesis, McGill University, Montreal, QC, Canada, 2017.
18. Wang, Y.; Gu, C.; Tang, Q.; Shi, Y.; Zhang, H.; Ye, L. Experimental study on seismic behavior of buckling-restrained steel plate shear wall panel element by cold-formed steel with hat-section under pure shear load. *J. Build. Struct.* **2020**, *41*, 49–57+64. (In Chinese)
19. Brière, V.; Santos, V.; Rogers, C.A. Cold-formed steel centre-sheathed (mid-ply) shear walls. *Soil Dyn. Earthq. Eng.* **2018**, *114*, 253–266. [CrossRef]
20. Yu, C.; Yu, G.; Wang, J. Optimization of Cold-Formed Steel Framed Shear Wall Sheathed with Corrugated Steel Sheets: Experiments and Dynamic Analysis. In Proceedings of the 2015 ASCE Structures Congress, Portland, OR, USA, 23–25 April 2015; pp. 1008–1020.
21. Zhang, W.; Mahdavian, M.; Li, Y.; Yu, C. Experiments and simulations of cold-formed steel wall assemblies using corrugated steel sheathing subjected to shear and gravity loads. *J. Struct. Eng.* **2016**, *143*, 04016193. [CrossRef]
22. Ngo, H.H. *Numerical and Experiment Studies of Wood Sheathed Cold-Formed Steel Framed Shear Walls*; Johns Hopkins University: Baltimore, MD, USA, 2014.
23. Xu, P. *Seismic Performance Investigation of New-Type Cold-Formed Steel Framed Sheer Walls with Steel Sheathing*; Southeast University: Nanjing, China, 2017. (In Chinese)
24. Yi, J.; Gil, H.; Youm, K.; Lee, H. Interactive shear buckling behavior of trapezoidally corrugated steel webs. *Eng. Struct.* **2008**, *30*, 1659–1666. [CrossRef]
25. *GB/T228.1-2010*; Metallic Materials-Tensile Testing-Part 1: Method of Test at Room Temperature. China Standards Press: Beijing, China, 2010. (In Chinese)
26. *AISI S100-16*; North American Specification for the Design of Cold-Formed Steel Structural Members. American Iron and Steel Institute: Washington, DC, USA, 2016.
27. Xie, Z.; Yan, W.; Yu, C.; Mu, T.; Song, L. Experimental investigation of cold-formed steel shear walls with self-piercing riveted connections. *Thin-Walled Struct.* **2018**, *131*, 1–15. [CrossRef]
28. Dassault System Simulia Corp. *ABAQUS Analysis User's Manual Version 6.10*; Dassault System Simulia Corp.: Providence, RI, USA, 2010.
29. Schafer, B.W.; Li, Z.; Moen, C.D. Computational modeling of cold-formed steel. *Thin-Walled Struct.* **2010**, *48*, 752–762. [CrossRef]
30. Chaboche, J.L. Time independent constitutive theories for cyclic plasticity. *Int. J. Plast.* **1986**, *2*, 149–188. [CrossRef]
31. Wang, M.; Shi, Y.; Wang, Y. Equivalent constitutive model of steel with degradation and damage. *J. Constr. Steel Res.* **2012**, *79*, 101–114. [CrossRef]
32. Shi, Y.; Wang, M.; Wang, Y. Experimental and constitutive model study of structural steel under cyclic loading. *J. Constr. Steel Res.* **2011**, *67*, 1185–1197. [CrossRef]
33. ABAQUS. *ABAQUS/Standard User's Manual. Version 2019*; Dassault Systemes Simulia Corp.: Johnston, RI, USA, 2019.
34. *JGJ 227-2011*; Technical Specification for Construction of Low-Rise Cold-Formed Thin-Walled Steel Buildings. China Architecture and Building Press: Beijing, China, 2011. (In Chinese)
35. *JGJ/T 421-2018*; Technical Standard for Cold-Formed Thin-Walled Steel Multi-Storey Residential Buildings. China Architecture and Building Press: Beijing, China, 2018. (In Chinese)
36. *AISI S400-20*; North American Standard for Seismic Design of Cold-Formed Steel Structural Systems. American Iron and Steel Institute: Washington, DC, USA, 2020.

buildings

MDPI

Article

Simple Aseismic Reinforcement of Steel Structures Using Knee Braces with High-Hardness Vises

Hiroyuki Nakahara *, Ding Nan and Iathong Chan

Graduate School of Engineering, Nagasaki University, Nagasaki 852-8521, Japan;
bb52222101@ms.nagasaki-u.ac.jp (D.N.); chan.iathong@nagasaki-u.ac.jp (I.C.)
* Correspondence: nakaharahiroyuki@nagasaki-u.ac.jp; Tel.: +81-819-2895

Abstract: A novel technique for upgrading the seismic resistance of steel buildings by adding knee braces to existing structures using vises was proposed by researchers in 2022. A feature of this retrofitting method is the easy setup owing to its use of vises made from high-hardness metal. Tests were conducted to investigate two main failure modes: slipping failure at the connection and yielding and buckling failure of the knee brace. The retrofitting design is discussed based on a comparison between the slipping strengths obtained through tests and calculations. Furthermore, an analytical study, using the finite element method (FEM), was conducted to evaluate the test results of retrofitted frames that failed in terms of the yielding and buckling of the knee braces. The findings of the analyses are consistent with the test results. This study included a stress relaxation test to assess the long-term performance of the vises.

Keywords: seismic retrofit; FEM analysis; yielding strength; buckling strength; long-term performance

1. Introduction

In Japan, many buildings have been damaged by large-scale earthquakes. In one steel building, the beam–column connection was broken, as shown in Figure 1 [1]. This fracture caused critical damage not only to the structure but also to human lives. The number of fractures observed in old buildings with defects in welding is decreasing. Seismic retrofitting in existing buildings is an effective method to reduce the damage caused by earthquakes. Many important studies have been conducted on seismic retrofits by researchers worldwide. For example, Fang [2] submitted a report on the practical applications of seismic retrofitting in the past decade, while MacRae [3] detailed diverse brace structures. The energy dissipation capacities of steel dampers were explored in studies by Domenico et al. [4], Kishiki et al. [5], and Tamai et al. [6] by investigating plastic deformation behaviors.

The authors of this study suggest a new technique for improving the seismic stability of steel structures by employing knee braces that are connected to existing structures through vises, which are composed of a body and a screw bolt made of high-strength metal. This method is specifically designed for moment-resisting frames in steel buildings.

Figure 2 illustrates a bending moment diagram for a portal frame subjected to a horizontal force. In a simple frame, the maximum bending moments are typically observed at the ends of both the beam and column. Defects and imperfections in welding joints between beams and columns have been found to result from damage caused by earthquakes. Knee braces are added to the frame, as shown in Figure 3, which decrease the bending moments at the beam–column connections. Knee-braced moment-resisting frames were studied in previous research [7–14], which reported that the structural stiffness and energy dissipation of these frames were enhanced compared those of normal moment-resisting frames with equivalent structural members. Previous studies conducted by Harada [15] and Yamada [16] investigated seismic retrofitting methods that use knee braces. Harada's research focused on bolted joints between retrofitted frames and knee braces, whereas Yamada et al. studied the use of knee brace dampers for retrofitting over-track buildings.

Figure 1. Steel structure damaged by Kobe earthquake in 1995 (reprinted from Ref. [1]).

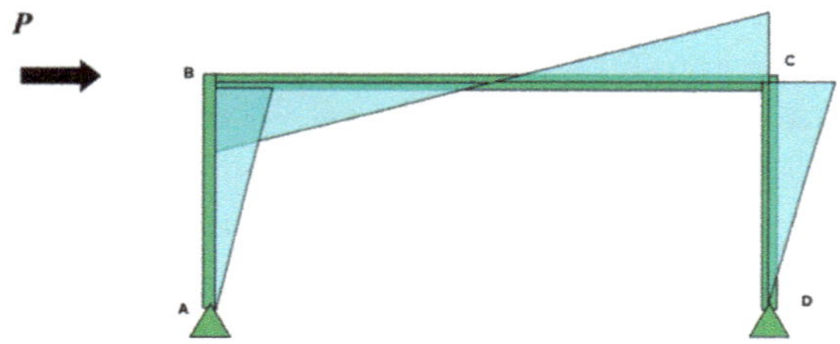

Figure 2. Bending moment diagram of non-retrofitted steel structure.

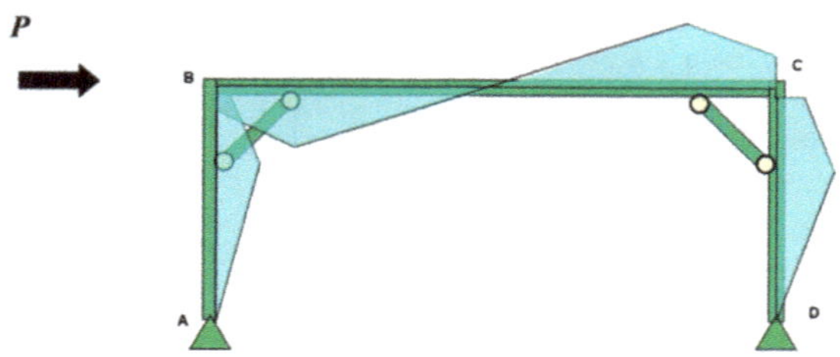

Figure 3. Bending moment diagram of retrofitted steel structure.

Knee braces serve the dual purpose of decreasing bending moments at the ends of structural members and enhancing the rigidity of beam–column connections. They are effective in preventing damage to the beam–column connections, but their implementation in existing buildings is challenging. The authors of this study suggest using the vise shown in Figure 4 as a potential solution.

The vise is designed to clamp the endplates of the knee brace onto existing beams and columns without requiring any welding or drilling for bolts. The bolthead of the vise is made from high-hardness metal. The loosening of the bolt is prevented because the head of the bolt bites into the steel plate when an initial torque is introduced.

Tests were conducted to investigate two main failure modes: slipping failure at the connection and yielding and buckling failure of the knee brace. The former was the test for the connecting design by the vises. The latter was conducted to estimate the energy dissipation performance of the knee braces. A simple equation for the slipping failure was proposed based on the test results reported in our previous paper [17]. The failure at the connections is normally avoided in seismic design. When the slipping strength is designed to be safe for the failure of the connections, the knee brace exhibits yield failure under tensile force and buckling failure under compressive force. The load versus deformation relations of the knee braces were obtained from the tests. A nonlinear finite element method (FEM) analysis was performed to assess the yielding and buckling behaviors of the knee brace. The analytical results obtained through FEM were compared with the experimental results reported in this paper. The FEM analysis was also conducted to examine the stress states of the vise itself, as reported in Appendix A. Furthermore, the long-term performance of high-hardness vises was examined through relaxation tests, which were then compared to the relaxation tests of high-strength bolts mentioned in previous studies.

Figure 4. Vise used for the seismic retrofitting of steel structures.

In Section 2, the experimental studies of the retrofitted frame using knee braces are shown. In Section 3, an FEM analysis is conducted and the findings are compared with the experimental results. In Section 4, the results of the relaxation tests are presented. Section 5 summarizes the outcomes of the study.

2. Experimental Study of Retrofitted Frame Using Knee Braces

A study on the use of knee braces to retrofit partial frames was conducted and is described in Ref. [17]. Figure 5 shows the test setup. Figure 6 shows the knee braces used as specimens for the investigation of slipping failure at the connection, and Figure 7 shows the knee braces used for the investigation of yielding and buckling. The length L, thickness t, and height h of the specimens are listed in Table 1. Table 2 shows the material properties of the beam and column used in the tests, and Table 3 shows the material properties of the knee braces used in the tests. The notations in the tables are as follows: tf: thickness of the flange of the H-shaped steel; t: thickness of the steel plate; σy: yield strength; σu: tensile strength; εst: strain at the point of strain hardening.

Table 1. Dimensions of knee braces in retrofitted steel frame specimens.

	h (mm)	L (mm)	t (mm)	θ (°)
SL-45-1	50.3	450	21.8	45
SL-45-2	50.3	450	21.8	45
SL-45-3	50.3	450	21.8	45
SL-30	50.0	384	21.5	30
nSL-6	5.9	200	21.5	45
nSL-9	9.0	200	21.5	45
nSL-12	11.9	200	21.5	45
nSL-15	14.8	200	21.5	45

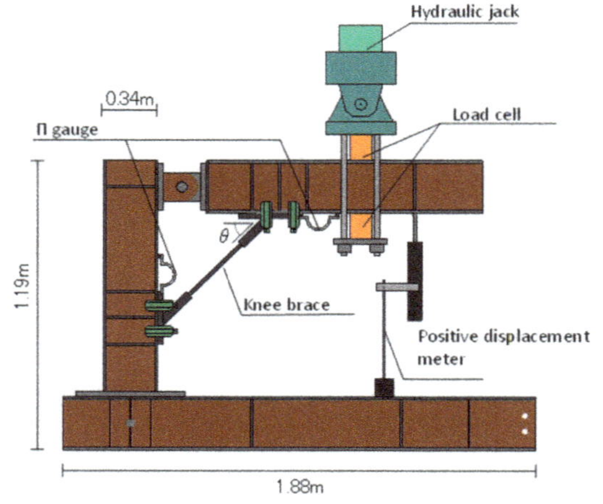

Figure 5. Test setup with loading and measuring devices (reprinted from Ref. [17]).

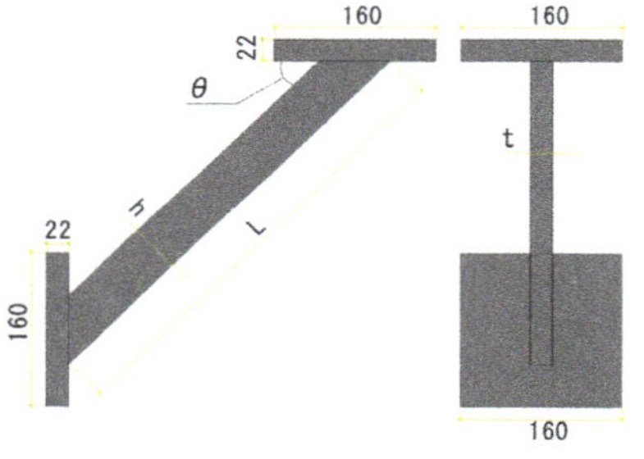

Figure 6. Shape of knee brace for investigating slipping failure at connection (unit: mm).

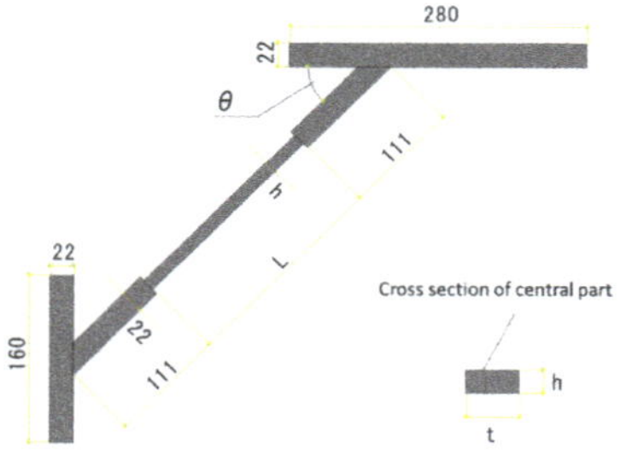

Figure 7. Shape of knee brace for investigating yielding at knee brace (unit: mm).

Table 2. Material properties of steel frame in test.

		tf (mm)	σy (N/mm^2)	σu (N/mm^2)	εst (%)
H-shaped Beam H200 × 200 × 6 × 9	SM490	8.9	392	520	2.83
H-shaped Column H200 × 180 × 6 × 9	H-SA700	9.0	795	862	—

Table 3. Material properties of steel plates for knee braces.

		t (mm)	σy (N/mm^2)	σu (N/mm^2)	εst (%)
SL-45	SM400	21.8	279	428	1.89
SL-30 nSL	SM400	21.5	271	434	1.85

The initial torque introduced using a torque wrench was 300 Nm for all vise bolts. The parameter for studying slipping behavior is the angle θ between the beams and knee braces. The angles of the specimens were 45° and 30°, respectively. The shapes of the specimens SL-45-1, SL-45-2, and SL-45-3 were the same. The slipping strength test results were validated against the evaluation method for the slipping strength, which was determined using Equation (1).

$$Q_A = \frac{nB\mu}{1+\mu\tan\theta}, \tag{1}$$

Equation (1) was derived in our previous paper [17], where n is the number of vises, B is the axial force of the bolt of the vise, μ is the frictional coefficient, and θ is the angle between the knee brace and the horizontal beam. When the values are n = 2, B = 75 kN, μ = 0.45, $\theta = \pi/4$, and $\tan\theta = 1$, then Q_A = 46.6 kN. When $\theta = \pi/6$, Q_A = 53.6 kN.

The load cells are set vertically along the axis of the hydraulic jack to measure the applied force, as shown in Figure 5. The slipping displacement is measured by π gauges laterally at the surface between the endplate of the knee brace and the steel beam. The strains on the surface of the knee brace along its axis are measured by four strain gauges glued to them. Axial deformations of the knee brace are measured using two displacement meters. The axial force N of the knee brace is obtained by multiplying the strain by its axial stiffness when the knee brace remains elastic. N can be determined in a rotational equilibrium state.

The experimental results for (a) SL-45-1, SL-45-2, SL-45-3, and (b) SL-30 are shown in Figure 8. The vertical axis shows the lateral component Q of N, and the horizontal axis shows the lateral displacement Δ_H, which is the average of the two displacements measured by the π gauges. The value of N is obtained from the strain gauge data shown in the figure. The slipping strengths are summarized in Table 4. The slipping strengths of the tests are defined as the forces at a slipping deformation of 0.5 mm, in accordance with AISC [18], and they exceed the slipping strengths, Q_A, calculated using Equation (1). The average strength of SL-45-1, SL-45-2, and SL-45-3 is 56.3 kN. The calculated value is 46.6 kN. The coefficient of variation is 4.5%. The calculation proposed in this study safely predicts the slipping strengths of the connections of all specimens. Using Equation (1), a retrofitting designer can choose the failure modes of the yielding and buckling of knee braces, which act as energy absorbers.

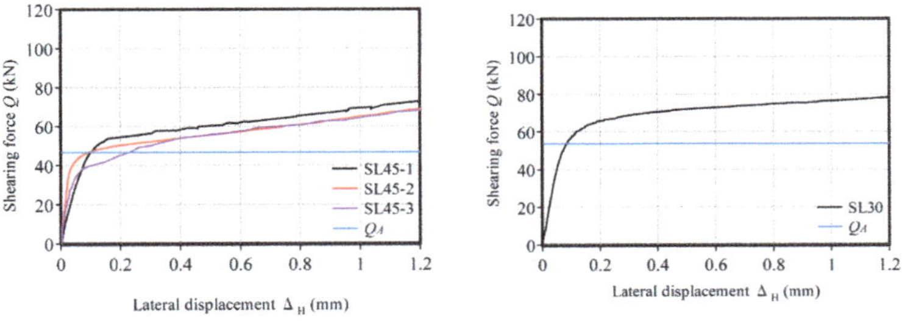

(a) (b)

Figure 8. Relationships between Q and Δ_H: (**a**) SL-45-1, SL-45-2, and SL-45-3; (**b**) SL-30.

The test parameter for observing the yield and buckling at the knee brace is the thickness, h, of the central part of the knee braces, 6, 9, 12, and 15 mm, corresponding to the names of the specimens: nSL-6, nSL-9, nSL-12, and nSL-15. The loading program is illustrated in Figure 9. The test results are discussed by comparing them with the analytical results, which are numerically calculated in Section 3.

Table 4. Test results for slipping strength.

	Test Results (kN)	Slipping Strength (kN)	
		Average	Calculated Value
SL-45-1	59.9		
SL-45-2	54.1	56.3	46.6
SL-45-3	55.0		
SL-30	71.9	71.9	53.6

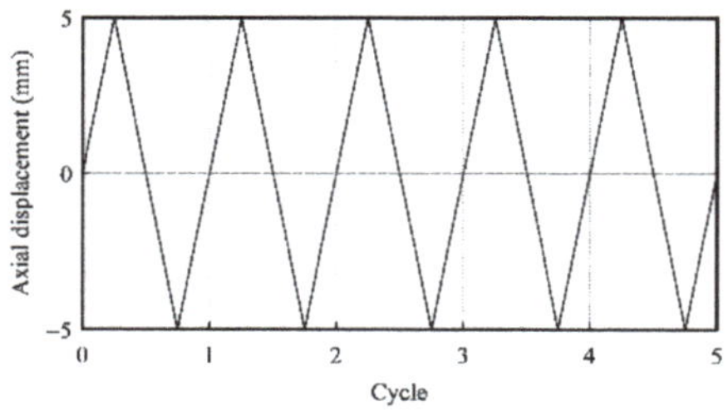

Figure 9. Loading program.

3. Analytical Study of Knee Brace and Discussion

An analytical study was conducted using the FEM program of Marc 2017.0.0., in which both material and geometric nonlinearities are considered. The main aim of the analytical investigation was to trace the load versus deformation relations of the knee braces obtained by the tests.

Figure 10 shows the analytical model of the knee brace. A solid element with ten nodes is used for the analytical model. An initial displacement, $L/1000$ (L is shown in Figure 7), is given in the orthogonal direction of the axis of the brace by considering that initial imperfections exist in the knee braces of the test specimens. The boundary conditions for the analysis are shown in Figure 11. The upper endplate is supported by rollers at four points, and the lower endplate is supported by pins at four points. The alternative displacements at the top of the knee brace are controlled to trace the load versus deformation relation of the test. The yield criterion follows von Mises stress, as shown in Figure 12.

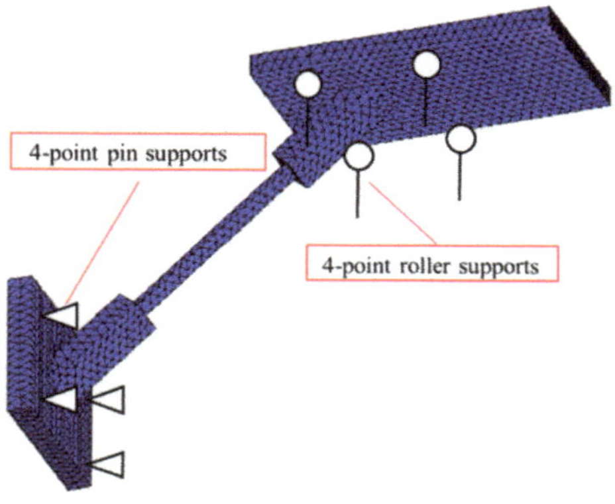

Figure 10. Element mesh and boundary conditions of analytical model.

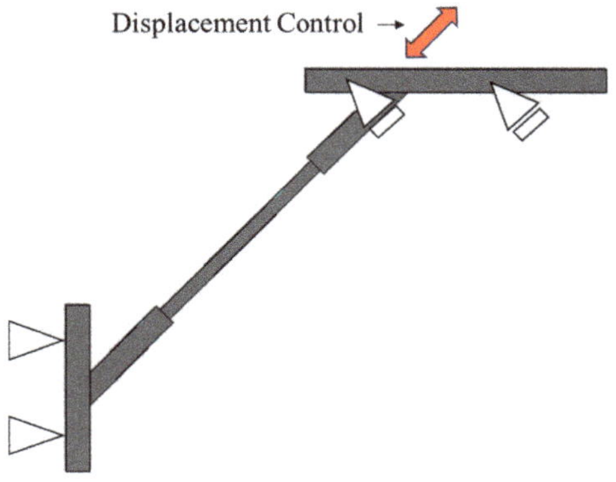

Figure 11. Boundary conditions and applied displacements of analytical study.

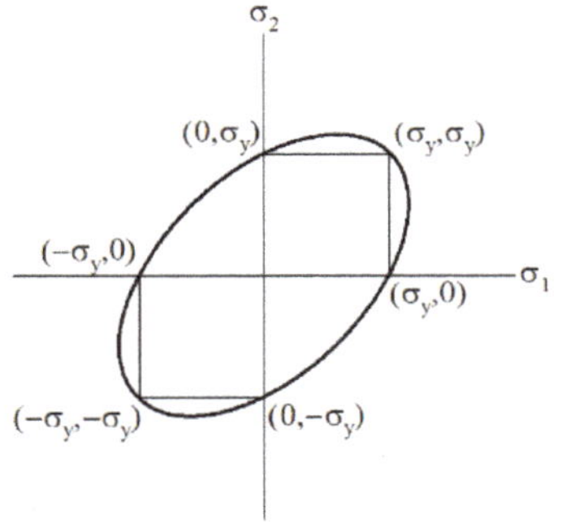

Figure 12. von Mises yield criterion.

Figure 13 shows the analytical results of the deformation and stress states using a contour diagram of the von Mises stress. The yellow and red areas indicate greater stress than the blue area. Figure 14 shows the buckling deformation during the test. The buckling deformations obtained from the test and the analysis show good agreement. Figure 15 shows the axial force, N, versus the axial displacement, Δ_A, relations of the knee braces for the experiments and analyses of nSL-6, 9, 12, and 15. The black and red lines in the figures represent the experimental and analytical results, respectively.

Figure 13. Deformation and stress contour analysis.

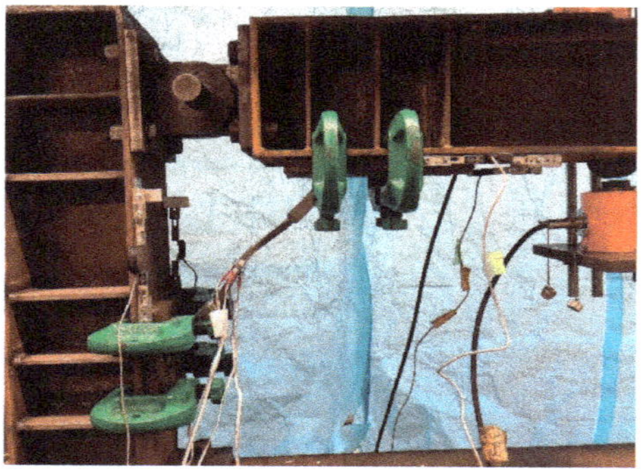

Figure 14. Buckling deformation test.

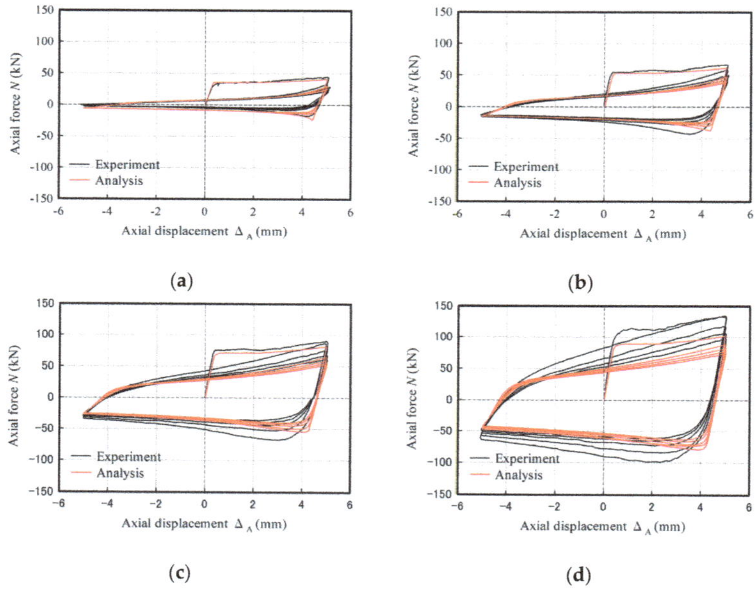

Figure 15. Relationships between N and Δ_A: (**a**) nSL6, (**b**) nSL-9, (**c**) nSL-12, and (**d**) nSL-15.

The values of N increase after yielding of the braces under tensile force owing to the strain-hardening effect. On the other hand, the values of N decrease after the buckling of the braces under compressive force. The tensile and buckling strengths of the four specimens increase with an increase in brace thickness. The analytical results match well with the experimental results.

The maximum values of the experimental and analytical results are summarized in Table 5. In the case of nSL-15, the test result is higher than the analytical result. This is because the test specimen was welded twice at the ends of the brace owing to the welding defects. The hardening effect caused by reheating is the reason why the test is stronger than the analysis. The yield strength of the test is over 6% of that of the analysis, except for that of nSL-15. The buckling strength of the analysis is smaller than that of the test results, except for nSL-6. This is because the initial displacement is assumed to be $L/1000$, which is too large for nSL-9 and nSL-12.

Table 5. Maximum values of experiment and analysis.

	Yield Strength (kN)			Buckling Strength (kN)		
	Exp.	Ana.	Exp./Ana. (%)	Exp.	Ana.	Exp./Ana. (%)
nSL-6	44.3	41.0	108	17.4	23.5	74
nSL-9	66.6	61.8	108	43.1	37.1	116
nSL-12	90.0	81.6	110	67.9	54.2	125
nSL-15	134	102	132	99.1	79.0	125

In our experimental and analytical results, the load versus deformation relationships show hysteretic loops that dissipate the input energy by their plastic behavior. The knee braces connected by the vises can exhibit effective energy absorption capacities, as shown in the results of recent studies [7–14].

After the tests, the buckled braces can be removed easily. The advantage of this method is not only the good attachment but also the detachment of the retrofitting members. Thus, this method can improve the resiliency of the building after large earthquakes due to its ability to replace the retrofitting devices.

4. Relaxation Test of Vise and Discussion

The long-term performance of joints that use bolts in steel structures has been widely studied through relaxation tests, including tests of high-strength bolts reported by Chesson [19] and Nah [20].

In Chesson's study [19], a stress relaxation test was conducted using high-strength A325 bolts. The short-term stress relaxation of these bolts introduced to a tension of 5000 pounds was examined. The experimental results showed that, after 1 min of tightening, the tensile stress in the bolts decreased by between 2% and 11%, with an average of approximately 5%. In Nah's study [20], relaxation tests were conducted on ASTM (American Society for Testing and Materials) A490 bolts [21]. The experimental parameter was the surface situation on tightened steel plates. The results showed that, after 1000 h, the retention rate was 86% for specimens without surface treatment and 81% for specimens with surface treatment.

Our proposal for joints applied using a vise is unique, as no previous study investigates the stress relaxation of the vise. Therefore, the long-term performance of the connection made using the vise was investigated through relaxation testing.

In the relaxation tests, two steel cuboids were prepared to measure the axial forces of the vise bolts. The dimensions of the cuboids were 44.8 mm and 44.4 mm in height, with cross sections of 27.6 × 26.9 mm and 27.5 × 26.4 mm, respectively. As shown in Figure 16, a steel plate with a thickness of 6 mm and the cuboid were clamped using the vise, and the axial force of the device was measured using strain gauges glued on four sides of the steel cuboid. Figure 17 shows the test setup, where the temperature was also measured using a thermocouple. The observations of strain and temperature were conducted over three months.

Figures 18 and 19 show the axial force versus the time relations of the two specimens. The vertical axis shows the axial force (kN), which was calculated by multiplying the average value of the four strain gauges by the axial stiffness of the cuboid. The horizontal axis shows the observation time. The initial axial forces were 75.2 kN and 77.7 kN for the two specimens, respectively. The average was 76.5 kN, close to the value of 75 kN, which is the nominal value of the initial force introduced by a torque of 300 Nm. The horizontal axis in Figure 18 shows the time (s) from 0 s to 300 s. The decreases in the axial forces of specimen 1 and specimen 2 were 2.6 kN and 1.7 kN, and the rates of decrease were 3.5% and 2.1%, respectively. No rapid drop in the axial force was observed for approximately five minutes after the introduction of the axial force. The decrease in the axial force was not as large as in the relaxation tests of high-strength bolt-joints reported by Chesson [19].

Figure 16. Photo of relaxation test.

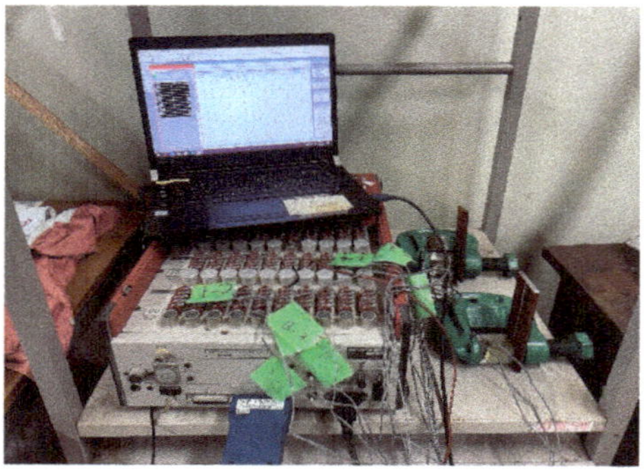

Figure 17. Test setup of relaxation experiment.

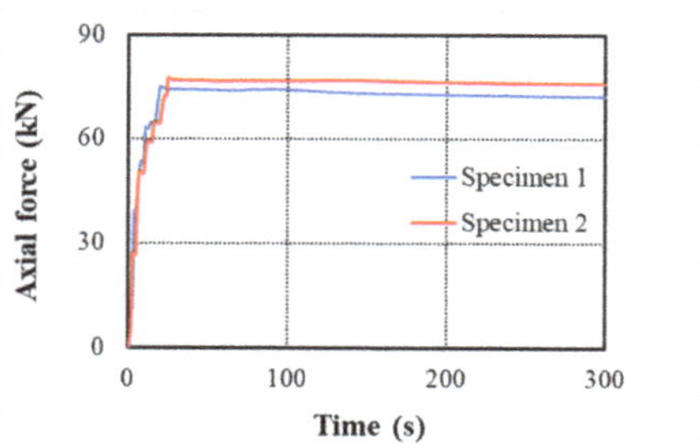

Figure 18. Axial force–time relationship over 300 s.

Figure 19 shows the axial force versus time (h) relationship recorded over three months, where the maximum value of the horizontal axis is 2200 h. The strain fluctuations of the test specimens have the same tendency as the temperature change observed by the thermocouple, and then the axial strain is corrected by considering a linear expansion coefficient of $10^{-5}/°C$. The retention rates of the two specimens were 92% and 94%, respectively. The stress relaxation tests showed the long-term connecting performance of the vise, in which the clipping force was maintained at over 90% of the initial force over three months. This result shows the same level of performance as the stress relaxation test of joints with high-strength bolts described by Nah [20].

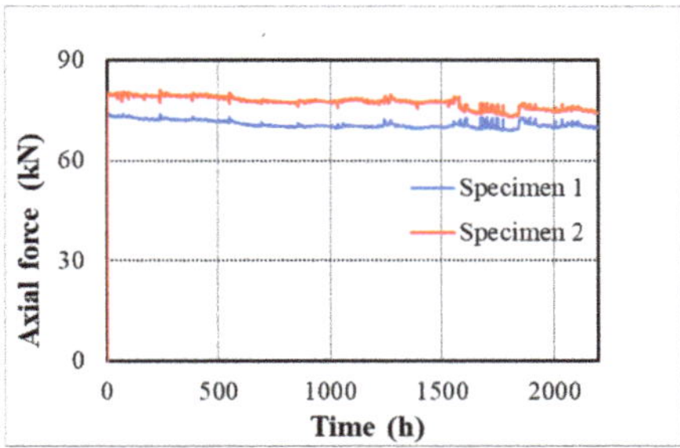

Figure 19. Axial force–time relationship over 2200 h.

5. Conclusions

In this study, a novel approach is proposed involving the use of knee braces attached to high-hardness vises. This is only applicable to open-frame steel structures. A feature of the retrofitting method is its easy setup owing to the use of vises. The boltheads of the vises, which are made from high-hardness metal, grip tightly between the steel plates of the connections. Testing and analysis were conducted to investigate the structural properties of the proposed method. Two failure modes were experimentally observed through the testing of specimens retrofitted using the knee braces: the slipping behavior at the connection and the yielding and bucking behavior at the knee brace.

The slipping strength between the endplate of the knee brace and the existing steel beam was investigated, and the slipping strengths at different angles were safely estimated using the calculating method proposed in our previous paper.

Nonlinear FEM analysis was conducted to evaluate the energy absorption performance of the knee braces. The characteristics of the load versus deformation relationship obtained through the testing of the retrofitted frames matched well with the FEM analysis results. The strengths of the yielding and buckling of the knee braces and the energy dissipating capacities of this retrofit were estimated using the FEM analysis. The knee braces connected by the vises showed the same level of energy dissipating capacity as those reported in previous studies.

Additionally, long-term relaxation tests were performed to investigate the practical use of the proposed method. The vises maintained their clipping forces at over 90% of the initial forces in the stress relaxation tests over 2200 h. No rapid loss in the clipping force of the vises was observed during the first five minutes of the relaxation test. The long-term performance of the vises was the same as that of high-strength bolts reported in previous studies.

The new seismic retrofitting method using vises made from high-hardness metal provides an easy construction technique with efficient seismic performance.

Author Contributions: H.N. designed the study. H.N., D.N. and I.C. were involved in analytical work, calculations, and data interpretation. All authors critically revised the manuscript, commented on the drafts of the manuscript, and approved the final paper. All authors have read and agreed to the published version of the manuscript.

Funding: This research received no external funding.

Data Availability Statement: The data that support the findings of this study are available from the corresponding author upon reasonable request.

Acknowledgments: The authors wish to express their gratitude to Touki Kawano, a student at Nagasaki University, and Kenta Okamoto, a technician at Nagasaki University, for their contributions to the experimental study.

Conflicts of Interest: The authors declare no conflicts of interest.

Appendix A

To ensure the seismic performance of steel structures joined using vises, it is important to confirm that the vises are not damaged from practical use. The mechanical performance of the vises was investigated using FEM analysis in Marc software. The dimensions of the vise are shown in Figure A1, and Tables A1–A3 summarize the chemical composition and mechanical properties of the vise.

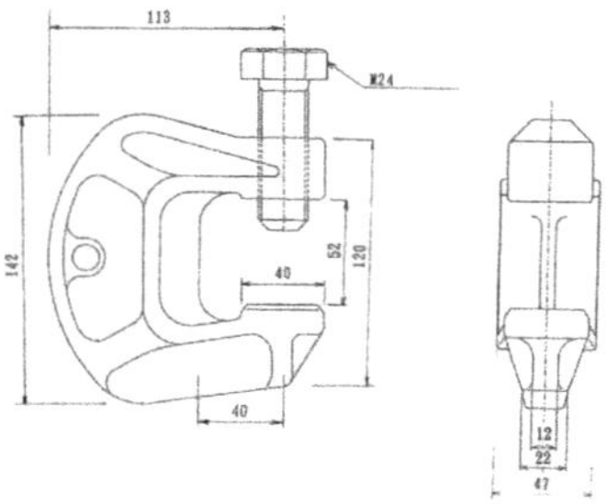

Figure A1. Dimensions of vise with high-hardness screw bolt.

Table A1. Chemical composition of vise (%).

	C	Si	Mn	P	S	Cu	Ni	Cr	Mo
Standard (%)	0.37~ 0.44	0.15~ 0.35	0.55~ 0.95	≤0.03	≤0.03	≤0.30	≤0.25	0.85~ 1.25	0.15~ 0.35
Measured (%)	0.41	0.26	0.7	0.019	0.015	0.11	0.08	1.11	0.17

Table A2. Chemical composition of bolt (%).

	C	Si	Mn	P	S	Cu	Ni	Cr	Mo
Standard (%)	0.37~ 0.44	0.15~ 0.35	0.55~ 0.95	≤0.03	≤0.03	≤0.30	≤0.25	0.85~ 1.25	0.15~ 0.35
Measured (%)	0.42	0.29	0.8	0.023	0.014	0.1	0.06	1.01	0.18

Table A3. Mechanical properties (JIS standard: reference value).

Spec	Tensile Test (No. 4 Test Piece)				Hardness Test
	σy (N/mm^2)	σu (N/mm^2)	Elongation (%)	Contraction (%)	Hardness (HB)
SCM440	834	980.7	12 more	45 more	285~352

There are two different loadings of the analysis, as shown in Figure A2. A force corresponding to shearing deformation, shown in Figure A2 (a), is applied if slipping occurs between the two steel plates. A force corresponding to tensile deformation, shown in Figure A2 (b), is applied when the axial force of the bolt is introduced by a torque wrench. Under these loading conditions, the stress state and load versus deformation relationships are discussed through numerical analysis.

The main body of the vise is analyzed by a mesh model with 10 nodes of solid elements, as shown in Figure A3. As shown in Figure A4, analytical supports are set at the top and bottom of the open part of the C shape of the vise. The four points at the bottom of the open part are fixed by pin supports. A horizontal sliding support is applied in the analysis of (a), and a vertical sliding support is applied in the analysis of (b). When the displacements are applied at the top of the vise, the reaction and relative displacements are obtained in each direction. The von Mises yield condition is used. The stress–strain model is defined as a bilinear elastoplastic model in Figure A5, and the material properties are listed in Table A4. The strain-hardening coefficient is the inclination of the line between the yielding point and the tensile strength corresponding to a strain of 20%. The relationship between true stress and logarithmic strain is applied to the analysis.

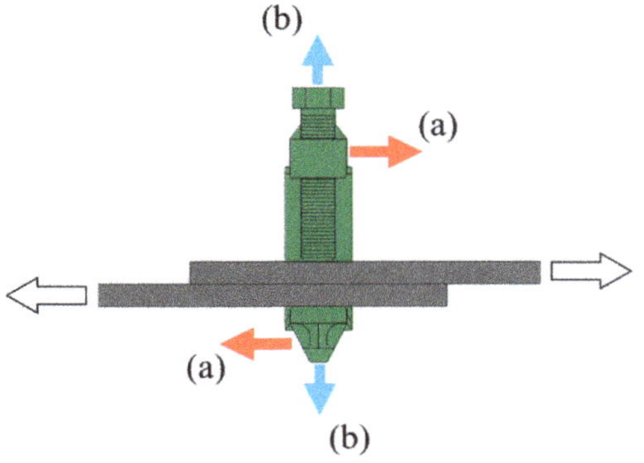

Figure A2. Applied forces to vise: (a) slipping force; (b) axial force.

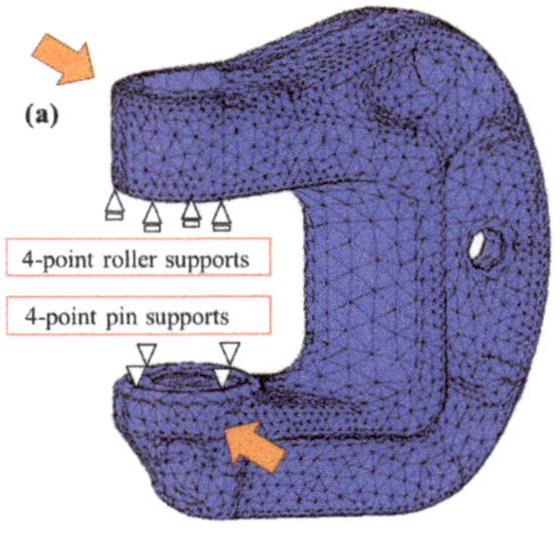

Figure A3. Boundary condition and direction of applied force in FEM mesh model of vise.

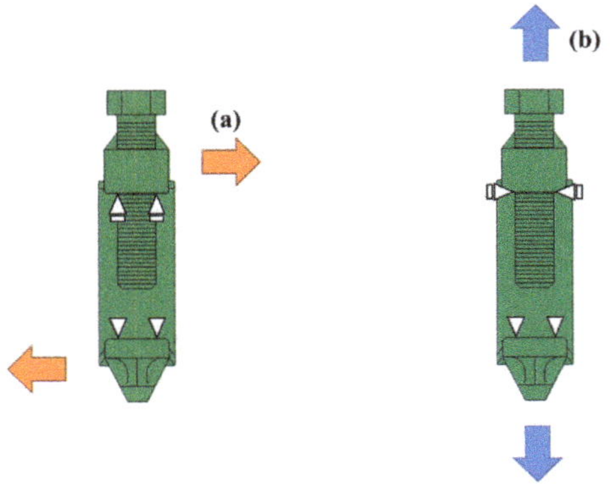

Figure A4. Boundary condition and direction of force in vises: (a) slipping force; (b) axial force.

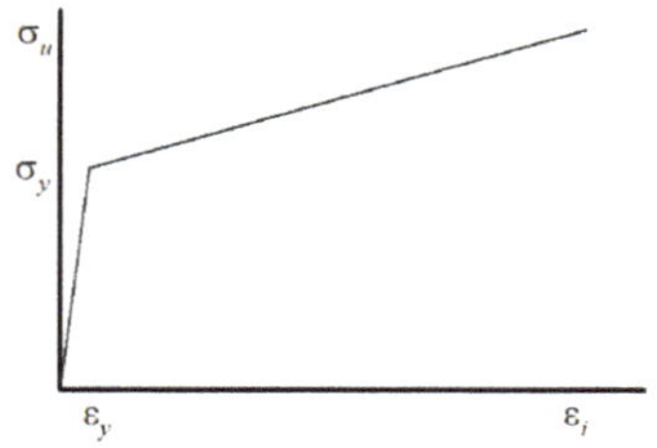

Figure A5. Bilinear stress–strain model.

Table A4. Constants in the material model.

σy (N/mm^2)	σu (N/mm^2)	εt (%)
834	980.7	20

σy: yield strength; σu: tensile strength; εt: strain at the tensile stress.

Figure A6 presents the load–displacement relationship in the direction of (a). In the figure, softening behavior is observed at approximately 30 kN; however, after that, the strength increases almost linearly up to 80 kN. After the steel plates joined by the vises start to slip, the bolt sustains a horizontal force. In this case, the main body of the vise, which provides twisting deformation, is subjected to the shearing force shown in Figure A2 (a) and Figure A4 (a).

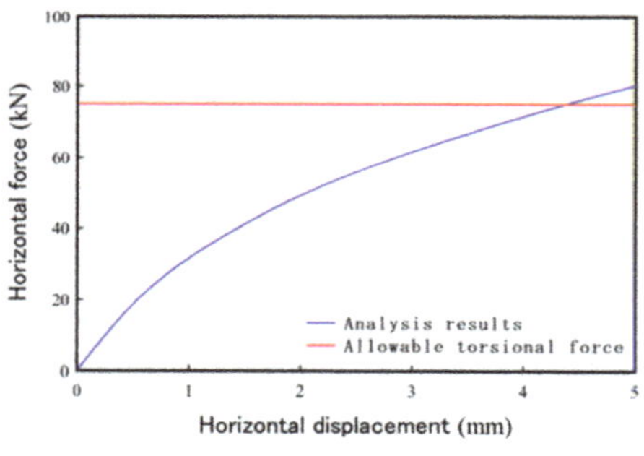

Figure A6. Load–displacement relationship under force applied in direction of (a).

In the study of the vise itself, an allowable force of shear is set as 75 kN per vise. The force of the screw bolt of the vise in the direction orthogonal to the bolt was defined as the "digging strength", which was described in a study conducted by Uno [22]. The digging strength corresponds to the axial force of the bolts of 75 kN, since the shape of the bolthead is triangular at 45-90-45.

As shown in Figure A6, even when the shearing force reaches 75 kN, the vise maintains its strength. Figure A7 presents a contour diagram of von Mises stress at a displacement of 4 mm in case (a). In the figure, some parts of the vise are yielding; however, the stresses in most parts remain elastic. This indicates that no practical problems arise when using a vise for seismic retrofitting.

Figure A8 shows the load–displacement relationship in the direction of (b). The opening deformation of the vice is caused by a reaction from the axial force of the screw bolts introduced from a torque wrench. In the figure, the opening force remains elastic up to 200 kN. Thus, an allowable axial force of the bolt is set as 200 kN. This is shown by the solid red line in Figure A8. Since the compressive force introduced into the bolt is 75 kN (red dotted line in Figure A8) during normal use of the vise, there is a sufficient margin in terms of strength.

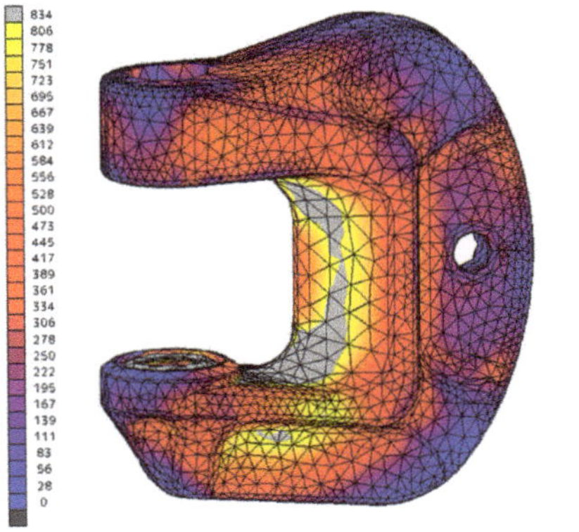

Figure A7. Contour diagram of von Mises stress at a displacement of 4 mm in case of (a).

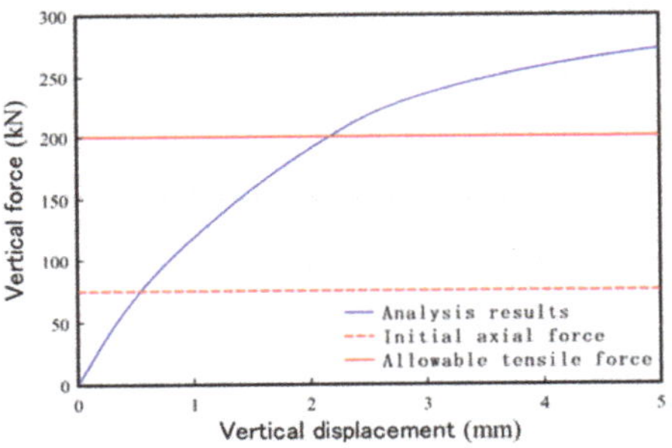

Figure A8. Load–displacement relationship under force applied in direction of (b).

Figure A9 shows a contour diagram of the von Mises stress at a displacement of 2 mm in case (b). In this figure, the yielding parts are displayed along the inside of the vise; however, the stresses of the other parts remain under yielding stress at a load of 200 kN. The applied load is 75 kN under normal use. Even if a double load is applied, the vise retains its connecting performance.

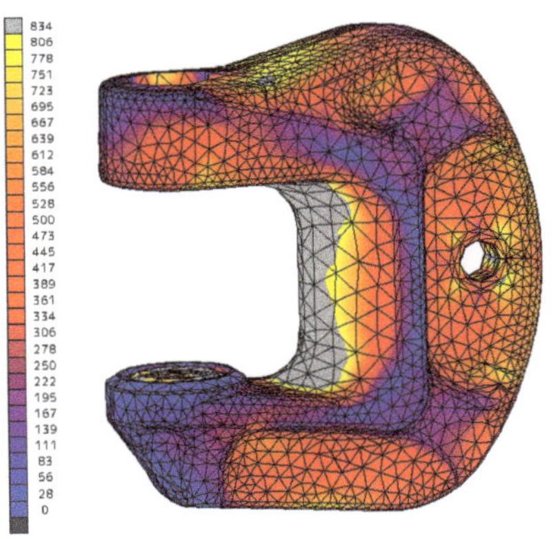

Figure A9. Contour diagram of von Mises stress at a displacement of 2 mm in the case of (b).

References

1. Steel Committee of Kinki Branch the Architectural Institute of Japan (AIJ). *Reconnaissance Report on Damage to Steel Building Structures Observed from the 1995 Hyogoken-Nanbu (Hanshin/Awaji) Earthquake*; Steel Committee of Kinki Branch the Architectural Institute of Japan (AIJ): Kyoto, Japan, 1995.
2. Fang, C.; Wang, W.; Qiu, C.; Hu, S.; MacRea, G.A.; Eatherton, M.R. Seismic resilient steel structures: A review of research, practice, challenges and opportunities. *J. Constr. Steel Res.* **2022**, *191*, 107172. [CrossRef]
3. MacRae, G. Low Damage Design of Steel Structures. Steel Innovations 2013 Workshop, Christchurch, New Zealand. Available online: https://www.scnz.org/wp-content/uploads/2020/11/Low-Damage-Design-of-Steel-Structures_MacRae_Clifton.pdf (accessed on 3 July 2024).
4. Domenico, D.; Ricciardi, G.; Takewaki, I. Design strategies of viscous dampers for seismic protection of building structures: A review. *Soil Dyn. Earthq. Eng.* **2019**, *118*, 144–165. [CrossRef]
5. Kishiki, S.; Uehara, D.; Yamada, S.; Suzuki, K.; Saeki, E.; Wada, A. Behavior of beam splices with energy dissipating elements at the bottom flange. *J. Struct. Constr. Eng. AIJ* **2005**, *70*, 135–143. [CrossRef] [PubMed]
6. Tamai, H.; Takamatsu, T.; Murata, M.; Kadoya, H. On Hystereric dampers for steel buildings with bent steel plates (Part1 Loading tests on bent steel plate). *Bull. Hiroshima Inst. Technol.* **2010**, *44*, 151–156.
7. Shin, J.; Lee, K.; Jeong, S.H.; Lee, H.S.; Kim, J.K. Experimental and analytical studies on buckling-restrained knee bracing systems with channel sections. *Int. J. Steel Struct.* **2012**, *12*, 93–106. [CrossRef]
8. Hsu, H.; Juang, J.; Chou, C. Experimental evaluation on the seismic performance of steel knee braced frame structures with energy dissipation mechanism. *Steel Compos. Struct.* **2011**, *11*, 77–91. [CrossRef]
9. Zhang, L.; Tong, G.S. Out-of-plane stability of simply supported beams with knee braces. *J. Constr. Steel Res.* **2007**, *63*, 175–181. [CrossRef]
10. Clement, D.E.; Williams, M.S. Seismic design and analysis of a knee braced frame building. *J. Earthq. Eng.* **2004**, *8*, 523–543. [CrossRef]
11. Kim, J.K.; Seo, Y.G. Seismic design of steel structures with buckling-restrained knee braces. *J. Constr. Steel Res.* **2003**, *59*, 1477–1497. [CrossRef]
12. Hsu, H.-L.; Li, Z.-C. Seismic performance of steel frames with controlled buckling mechanisms in knee braces. *J. Constr. Steel Res.* **2015**, *107*, 50–60. [CrossRef]
13. Honma, S.; Ebato, K.; Harada, Y. Ductile steel knee brace with built-in comb-shaped seismic damper. *J. Constr. Steel Res.* **2021**, *184*, 106765. [CrossRef]
14. Li, R.; Yuan, Z.; Qi, L.; Yuan, C.; Xue, J. Improvement effects of knee-bracing friction dampers on cyclic behavior of steel joints in antique buildings. *Structures* **2024**, *60*, 105919. [CrossRef]
15. Harada, Y.; Ebato, K.; Honma, S.; Takimoto, T. Experimental Study on Seismic Retrofit by Using Supplemental Knee Braces Attached to Steel Members with Semi-Rigid Bolted Connections. In Proceedings of the 15th World Conference on Earthquake Engineering, Lisbon, Portugal, 24–28 September 2012; Volume 5.
16. Yamada, S.; Shimizu, K.; Takei, Y. Seismic Retrofitting Effect of Reinforcing Over-Track Buildings with Knee-Brace Dampers. *Q. Rep. RTRI* **2011**, *52*, 217–223. [CrossRef]
17. Nakahara, H.; Ding, N.; Shimomura, T. Seismic retrofitting method for steel structures by knee braces jointed by high hardness vises. *J. Civ. Eng. Manag.* **2022**, *29*, 67–76. [CrossRef]
18. AISC-Research Council on Structural Connections. *Load and Resistance Factor Design Specification for Structural Joints Using ASTM A325 or A490 Bolts*; AISC-Research Council on Structural Connections: Chicago, IL, USA, 1994.
19. Chesson, E.; Munse, W.H. Studies of the behaviour of high strength bolts and bolted joints. In *Engineering Experiment Station Bulletin 469*; University of Illinois College of Engineering: Chicago, IL, USA, 1964.
20. Nah, H.S.; Lee, H.J.; Kim, K.S. Evaluation of Long-term Relaxation for High-strength Bolted Connections. In Proceedings of the ICAPP '10, San Diego, CA, USA, 13–17 June 2010.
21. AISC-Research Council on Structural Connections. *Specification for Structural Joints Using ASTM A325 or A490 Bolts*; AISC-Research Council on Structural Connections: Chicago, IL, USA, 2004.
22. Uno, N.; Inoue, K.; Shimura, Y.; Wakiyama, K. Coefficient of friction between steels with different hardness. *J. Struct. Constr. Eng. AIJ* **1997**, *494*, 123–128. [CrossRef] [PubMed]

buildings

Article

Test and Analysis for Shearing Behavior of Circular CFST Columns

Hiroyuki Nakahara [1,*], Kazuhiro Uchida [2] and Yuto Yanai [3]

1 Graduate School of Engineering, Nagasaki University, Nagasaki 852-8521, Japan
2 Faculty of Engineering, Kyushu Sangyo University, Fukuoka 813-8503, Japan; uchida.k@ip.kyusan-u.ac.jp
3 General Building Research Corporation of Japan (GBRC), Osaka 565-0873, Japan; yanai@gbrc.or.jp
* Correspondence: nakaharahiroyuki@nagasaki-u.ac.jp; Tel.: +81-819-2895

Abstract: Concrete-filled steel tubular (CFST) structures are well known to possess high strength and ductility. CFST members are used under complex stress states, such as beam–columns, piers, caissons, or other foundation components. Recommendations for the design and construction of concrete-filled steel tubular (CFST) structures were published in 1997 and revised in 2008 in Japan. In the recommendations, calculation methods for the axial strength and flexural strength of CFST columns were established on the basis of experimental results of more than 400 specimens; however, the test results of the columns that failed in shear referred to only 12 specimens in the recommendations. It is necessary to accumulate experimental data on the shear strengths and behaviors of CFST columns. Tests and analyses have been carried out on eight circular CFST column specimens with a shear span ratio of 0.75. The diameter-to-thickness ratio of the steel tube is approximately 34. The shearing capacities of the tests were underestimated by over 20% errors using the calculation method of the CFST Recommendations in Japan. The load versus deformation relations obtained by the tests were well traced by 3D-FEM analysis. The shearing capacities were estimated as an average of 12% errors using 3D-FEM analysis.

Keywords: short column; flexural strength; load–deformation relation; 3D-FEM

1. Introduction

Concrete-filled steel tubular (CFST) columns provide a number of advantages when used in seismic-resistant frames. The inner concrete increases the axial and flexural stiffness and load-carrying capacity while permitting more slender elements. Because slender beam–columns are normally predicted to fail in buckling or flexure, few studies of short columns of CFST that fail in shear have been conducted. It is necessary to provide a design formula for very short columns in structural designs based on a few specimens that fail in shear. To obtain data on the shearing failure of short CFST columns, we have experimentally investigated the load versus deformation characteristics of CFST specimens subjected to shearing force under constant axial compression. The test results were compared with those of the 3D-FEM analyses.

An extensive number of experimental studies for CFST members were conducted in the fifth phase of the U.S.–Japan Cooperative Earthquake Research Program from 1992 to 1996. The total summaries of this program were reported in papers by El-Tawil and Bracci [1] and by Goel [2]. The analytical model for composite and hybrid structures was overviewed by Spacone and EI-Tawil [3]. The test results for axial compression were examined in two papers by Varma [4] and the authors [5]. The beam-column tests were reported in Refs. [6,7]. The research on the connections is presented in the papers [8–11].

The authors proposed uniaxial stress–strain curve models for inner concrete and steel tubes of CFSTs in Ref. [5] in this program. The authors also investigated the flexural behavior of single-curvature columns through experimental and analytical studies [12]. In this analytical study, the proposed uniaxial stress–strain curve models were used to predict the bending behavior accurately as a multi-fibers model. One of the authors had already studied the triaxial state of the components of composite and hybrid structures in Refs. [13,14] by a finite elements method (FEM).

In Japan, "Recommendations for Design and Construction of Concrete Filled Steel Tubular Structures [15]" was published in 1997 and revised in 2008. The Recommendation includes the results of the U.S.–Japan research and summarizes more than 400 specimens to establish the design methods for axial and flexural capacities but refers to only 12 specimens to verify the calculation method for shearing capacity. The 12 test results of shear failure of CFST short columns, of which shear span ratios were from 1.0 to 1.5, were reported by Sakino [16]. The sectional shapes of the twelve specimens were limited to 100 mm × 100 mm squares. The concrete strength of the tests was around 20 MPa, which was relatively lower than that of normal use of CFST. This study focuses on the shearing capacity and behavior of a circular CFST column.

The characteristic behaviors of CFST members have been investigated by many researchers. Analytical studies on CFST columns under axial compression were conducted by Choi and Xiao [17] and Hu et al. [18]. There were some analytical studies on predicting the behavior of beam–columns by Varma et al. [19] and the behavior of beam–column connections by Chiew et al. [20]. The CFST structure has the feature of the mutual confining effects between inner concrete and steel tube. The confining effect was essentially explained by three-dimensional models. These previous studies showed that FEM analysis is an effective way to predict complex three-dimensional stress states, for example, the study by Tort and Hajjar [21]. The authors also conducted a numerical investigation by 3D-FEM analysis to trace the relations between the shearing force and drift angle of the circular CFST columns.

Important research on the shearing behavior of circular CFST members was reported by Roeder et al. [22]. In this study, a database of test results of circular CFST that failed in shear was built, and they proposed a new calculation method for the shearing capacities of circular CFST columns. The Roeders' study focused on beams or piles without axial force. Our study focuses on the columns with axial force.

In Section 2, an outline of the experimental work is presented. Section 3 shows the experimental results. A model for FEM analysis is shown in Section 4, and the analytical results are discussed in Section 5. In Section 6, the results of the test, analysis, and calculation are compared. Section 7 summarizes the outcomes of the study.

2. Specimen and Test Setup

A test matrix is presented in Table 1. A primary test parameter is the ratio of the axial force N to the nominal axial force capacity N_0. Annealed and non-annealed steel tubes are used. The average compressive strength of concrete cylinders is 50.9 MPa. The yield stress of the non-annealed steel tube is 531 MPa, and that of the annealed one is 507 MPa. The notations in the tables are as follows: D: diameter of steel tube, t: wall thickness of steel tube, D/t: diameter to thickness ratio, a/D: shear span to depth ratio, a: shear span, ${}_c\sigma_B$: strength of concrete cylinder, σ_y: yield stress of steel tube, N: constant axial force applied.

Table 1. Test matrix.

Specimen	D (mm)	t (mm)	D/t	a/D	${}_c\sigma_B$ (N/mm^2)	σ_y (N/mm^2)	N/N_0	Annealing
N75-1	165	4.87	34	0.75	50.8	531	0.1	× (Non-annealed)
Non							0.2	
N75-3							0.3	
N75-4							0.4	
A75-1						507	0.1	○ (Annealed)
A75-2							0.2	
A75-3							0.3	
A75-4							0.4	

The shape of the test specimen is shown in Figure 1. The mid-height region of the specimen is broken by the loading apparatus shown in Figure 2. The specimens are fabricated by welding steel tubes and plates. Four 19 mm thick plates are set laterally as loading stubs. Two central plates are carefully placed in the steel tube. The two plates are welded on one side of each plate, i.e., the upper and lower sides of the test part are not welded to avoid the influence of welding heat on the test specimen.

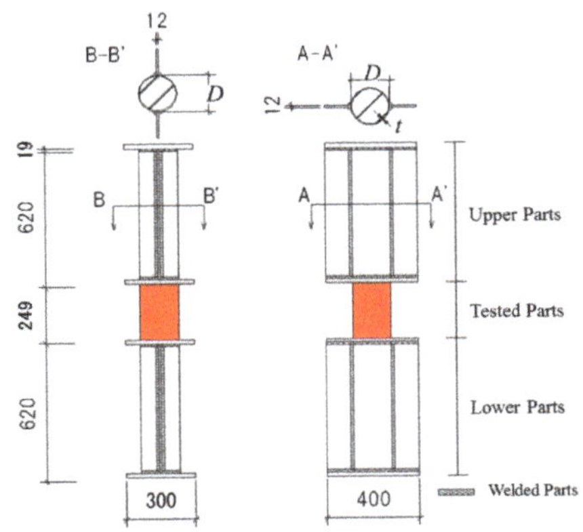

Figure 1. Test specimen.

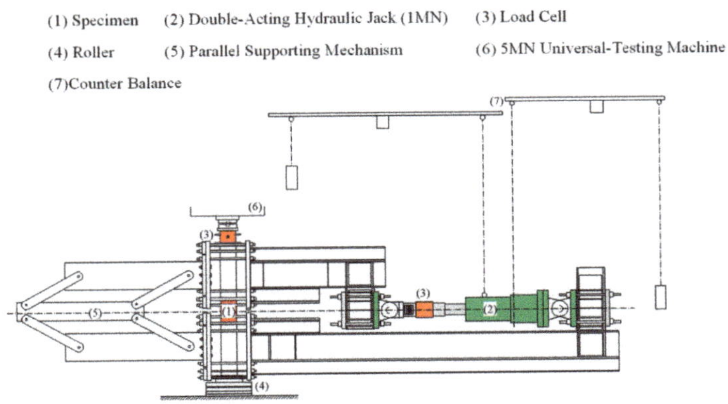

Figure 2. Loading apparatus.

The axial force N is adopted by the universal testing machine at Kyushu University. Lateral forces Q are applied by a double-acting hydraulic jack with a capacity of 1 MN. A bending moment diagram of the whole test setup is shown in Figure 3. The mid-height region of the specimen deforms anti-symmetrically, as shown in Figure 4.

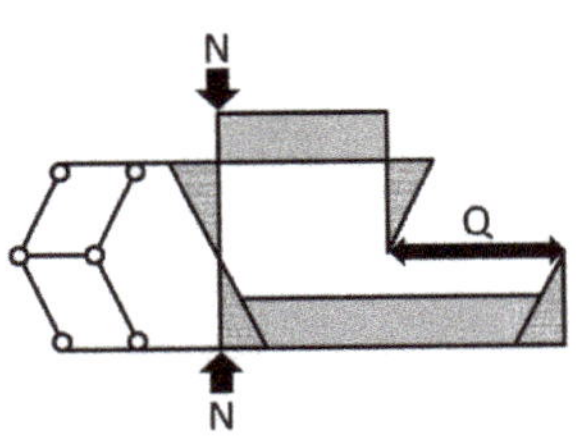

Figure 3. Bending moment diagram.

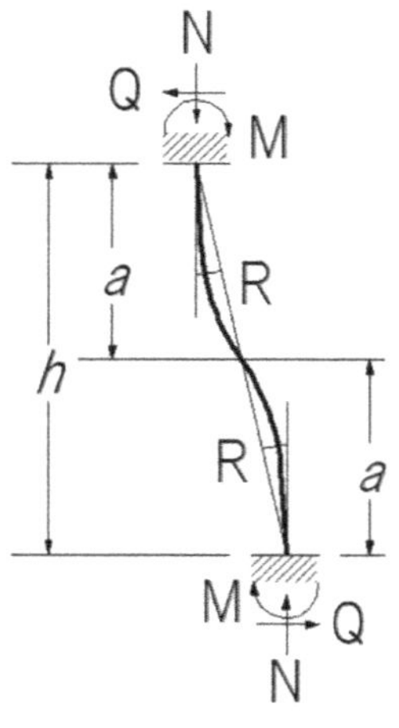

Figure 4. Deformation of tested part.

The deformation histories of the test are shown in Figure 5, where the vertical axis shows the drift angle *R* of the specimen, and the horizontal axis shows the number of loading cycles. The drift angles *R* are obtained by dividing lateral displacements by the clear height of the specimen. The test is controlled under *R* according to Figure 5, where the peak drift angle increases stepwise from 1.0/100 rad. to 4.0/100 rad. Three cyclic deformations are applied at each drift angle. When the specimens do not show maximum strength within *R* = 4.0/100 rad., extra deformations are applied. They are up to *R* = 8.0/100 rad. for non-annealed specimens and *R* = 6.0/100 rad. for annealed specimens.

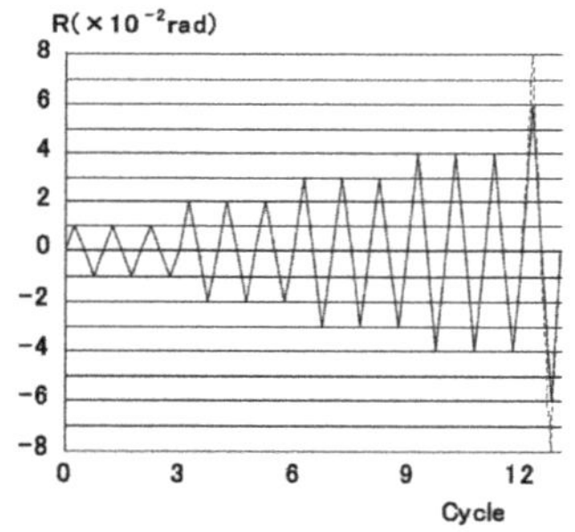

Figure 5. Deformation histories.

Figure 6 shows locations for transducers and strain gauges. The eight linear variable differential transformers (LVDTs) are used to measure the deformations of the specimens. The four of them are set horizontally on both sides of the specimen to measure the drift angle *R*. Axial displacements are obtained from four LVDTs set vertically at four corners of the stub of the steel plate. A total of ten strain gauges are used for each specimen. The gauges are available for the post-elastic range and are grouped into A and B. The gauges in group A glued on the flange longitudinally are used to measure the curvatures of the top and bottom of the specimen. If the strains in group A attain the yield strain, the specimen is judged to be yielding in flexure at this point. The three-axis gauges in group B, glued on the web, are used to determine two-dimensional stress tensors. From the stress tensor, von Mises stress σe is calculated by Equation (1).

$$\sigma_e{}^2 = \frac{1}{2}\left\{(\sigma_\theta - \sigma_z)^2 + \sigma_\theta{}^2 + \sigma_z{}^2 + 6\tau_{\theta z}{}^2\right\} \quad (1)$$

where $\sigma\theta$, σz are normal stresses of horizontal and vertical directions. $\tau\theta z$ is the shearing stress. They are obtained from the strain data on the web of specimens. If the value of σe attained the yield stress, the specimen is judged to be yielding in shear.

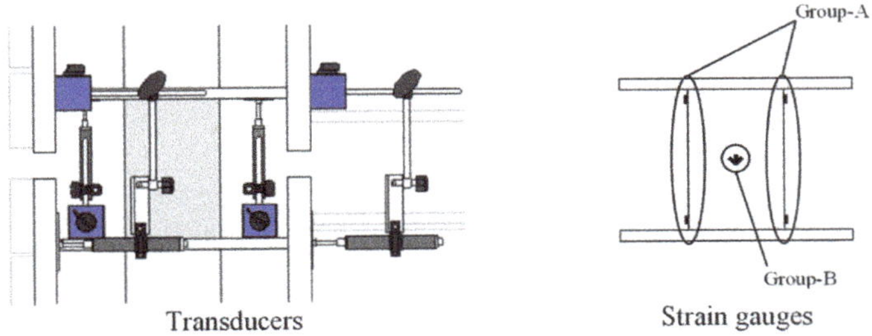

Figure 6. Locations of transducers and strain gauges.

3. Experimental Results

The relations between shearing force Q and drift angle R are shown in Figure 7. The symbol ○ is the coordinate (Qy, Ry), which indicates the occurrence of yielding in shear. The loading tests for N75-2, N75-3, A75-1, A75-2, and A75-3 yield in shear during the loading to R = 2.0/100 rad. before the observation of flexural yielding; then, the shearing forces of the specimens increase according to the increases of the displacements. N75-1 occurs shearing and flexural yielding at the same deformation. N75-4 and A75-4 attain their maximum strengths $Qmax$ around R = 3.0/100 rad., and then slight decreases in strengths are observed. Only N75-4 yields in flexure before the shear-yielding occurs.

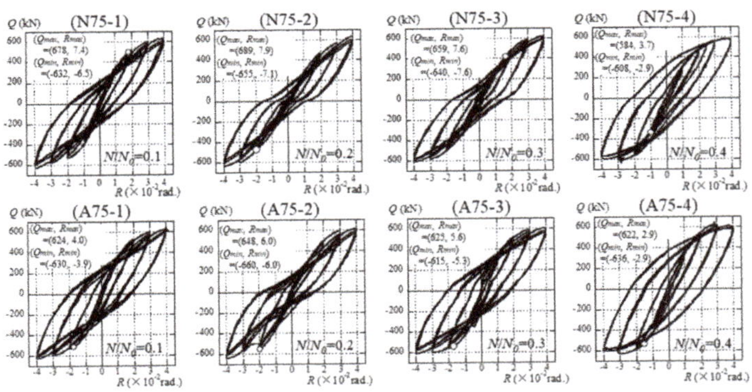

Figure 7. Relations between shearing force Q and drift angle R.

Figure 8 shows the pictures of the inner concrete of N75-1 and A75-1 after the tests. Both have clear diagonal cracks, which show typical shearing failure.

Figure 8. Cracks on inner concrete after test.

The test specimens do not have fractures or cracks at the welded parts or on the surface of the steel tubes. This indicates that the test specimens successfully exhibited shearing or flexural capacities. It is difficult to clearly distinguish the failure modes of shear and flexure. Comparisons between the shearing capacities and flexural capacities are discussed in Section 6, entitled "Comparisons of Test, Analysis and Calculation".

4. Finite Element Model for Analysis

An analytical investigation of the test specimens was carried out using finite element code DIANA version 9.5. The feature of the program includes the modified compressive theory for concrete [23]. The purpose of the analysis is mainly to estimate each of the shearing forces that the steel tube and the inner concrete sustained in the tests. The material properties of the concrete and steel are listed in Tables 2 and 3. The property of the interface element between the steel tube and the inner concrete is shown in Table 4. The concrete is modeled by solid elements. A total strain crack model based on the modified compression field theory is used for the nonlinear constitutive law. The tensile properties of the concrete are shown in Figure 9. The tensile strength is calculated by the expression appearing in "Design Guidelines for Earthquake Resistant Reinforced Concrete Building Based on Inelastic Displacement Concept" [24], published by AIJ. The Hordjik model [25] is used for the tension-softening function. The tensile fracture energy G_f is calculated by the expression used among "Standard Specifications for Concrete Structures" [26], published by JSCE. The compressive properties of the concrete are shown in Figure 10. The Feenstra model [27] is used to determine the compressive properties. The compressive fracture energy *Gc* is estimated by the expression of Nakamura and Higai [28]. The steel tube and the web of the stub are modeled by curved shell elements with four nodes. The flange of the stub is modeled by solid elements. A bilinear model, shown in Figure 11, is used for the material properties of the steel. The yield stress is determined by the von Mises yield criterion. The Coulomb friction model is used for the interface element between steel and concrete, as shown in Figure 12. The analytical properties of the interface elements used for the CFST column by Komuro et al. [29] are adopted.

Table 2. Material properties of concrete.

Specimen	E_c (N/mm^2)	${}_c\sigma_B$ (N/mm^2)	σ_t (N/mm^2)	G_c (N/mm)	G_f (N/mm)
N75 A75	3.12 × 104	50.9	2.35	62.6	0.108

Table 3. Material properties of steel.

Specimen	t (mm)	E_s (N/mm^2)	σ_y (N/mm^2)	H (N/mm^2)
N75 A75	4.87	2.01×10^5	531 507	2.01×10^3

Table 4. Properties of interface between steel tube and concrete.

Specimen	k_n (N/mm^3)	k_t (N/mm^3)	c (N/mm^2)	$tan\Phi$	f_t (N/mm^2)
N75 A75	1.0×10^4	1.0×10^3	0.783	3.6	0

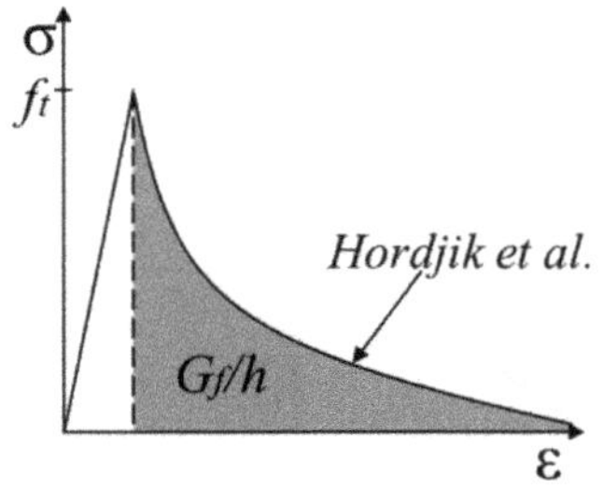

Figure 9. Tensile properties of concrete [25].

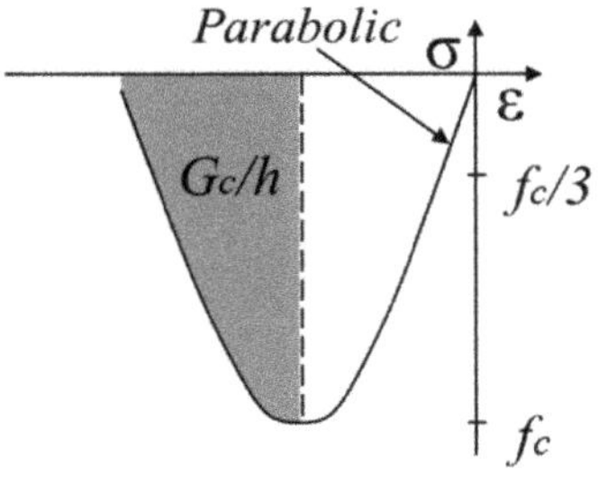

Figure 10. Compressive properties of concrete.

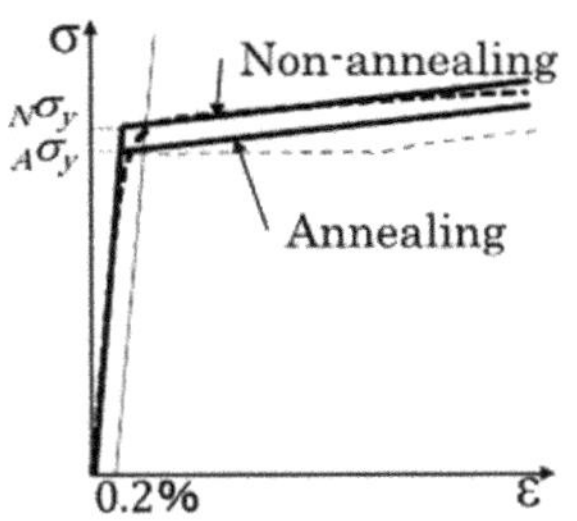

Figure 11. Material properties of steel.

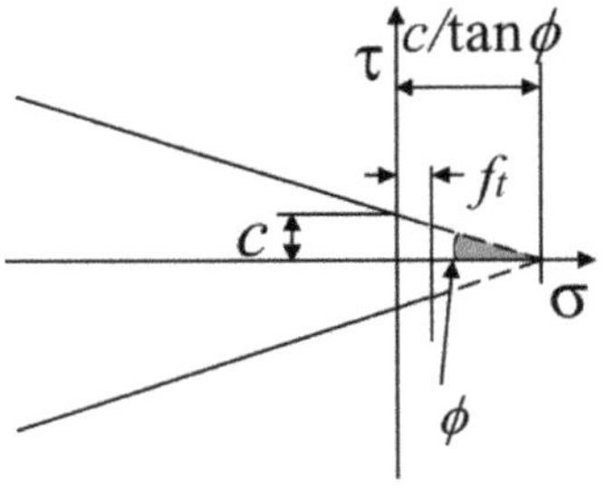

Figure 12. Coulomb–friction model.

The components of the FEM analysis are illustrated in Figure 13. Half of the entire specimen is investigated because the specimen is symmetrical in the Z–X plane. Spring elements are installed at the center section of the specimen to measure the shearing forces of the steel tube and the inner concrete directly.

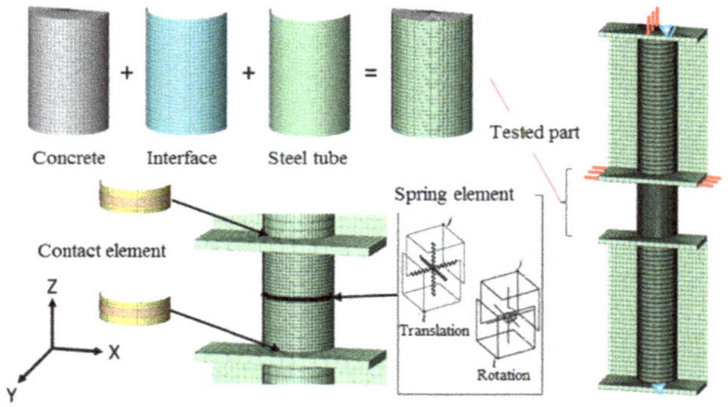

Figure 13. Finite element model for analysis.

A 0.4 mm clearance is provided between the steel tube and the flanges of the stubs because the diameter of the steel tube is 165.2 mm and the diameter of the hole of the flanges is 166 mm. There are gaps between the steel tube and flange of the stubs; hence, they are not welded at the sides of the test part but are welded at the sides of the stub parts. The authors intend to avoid the influence of welding heat on the critical section, as shown in Figure 1.

Contact elements are used in the clearance between the steel tube and the flange. Spring elements with high stiffness are used in the welded parts.

The axial force is applied at the top of the specimen. Lateral displacements are applied in opposite directions at the upper and lower flanges. The lateral displacements are given to the drift angle of $R = 4/100$ rad. in the X-axis direction. The drift angle R is obtained by dividing the relative displacements between the two flanges by the clear heights of the specimens. Displacements of all nodes on the Z–X plane are zero in the Y-direction. The central node at the top of the specimen is fixed in the X-direction. The central node at the bottom of the specimen is fixed in both the X and Z directions.

5. Analytical Results and Discussions

The relations between the shearing force Q and drift angle R are shown in Figure 14. The dotted lines show the experimental results, and the solid lines show the analytical results. The symbols △ and ○ indicate the occurrences of flexural and shearing cracks in the inner concrete in the analyses. The symbols ▲ and • indicate the flexural and shear yielding of the steel tub in the analyses. The symbol ▼ shows the analytical shear strength *Qana*.

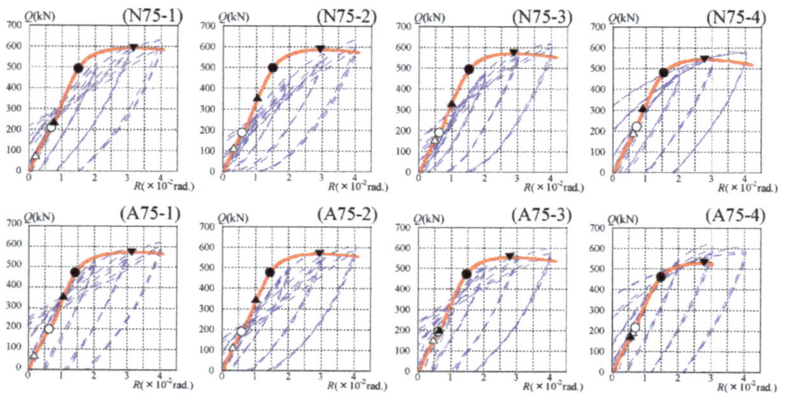

Figure 14. Relations between shearing force Q and drift angle R in the analysis.

With respect to the crack of the inner concrete, shearing cracks are observed from around $R = 0.6/100$ rad. in the analyses. The shearing cracks appear after the observation of the flexural cracks. The increase in the axial force ratio delays the drift angle when the flexural crack appears. With respect to the yielding of the steel tube, the shear yielding occurs around $R = 1.5/100$ rad. The flexural yielding occurs prior to the shear yielding. These results contradict experimental results. It is noteworthy that the maximum strengths are exhibited after the shear yielding of the steel tube in both the test and analysis.

The initial stiffness in the test and the analysis generally agree. The maximum strengths of the analyses are smaller than those of the tests for all specimens. During the tests, the shearing force of the specimens gradually increases after $R = 3/100$ rad. On the other hand, in the analysis, the shearing force decreases after the peak at $R = 3/100$ rad. The difference is due to the strain hardening of the steel tube. Because the test is carried out under cyclic loading, the strain hardening rate of the test is greater than that of the analysis under monotonic loading.

Contour diagrams of minimum principal stress of the inner concrete of the N75-1 and N75-4 specimens are shown in Figure 15, where the shape of the compression strut is confirmed. Furthermore, the increase in axial force enhances the shearing capacity because the compression region becomes larger. Contour diagrams of the shearing stress of the steel tubes of N75-1 and N75-4 specimens are shown in Figure 16, where the increase in axial force decreases the shearing stress of the steel tube. When the specimens exhibit maximum strength, the shear yielding of steel tubes is confirmed.

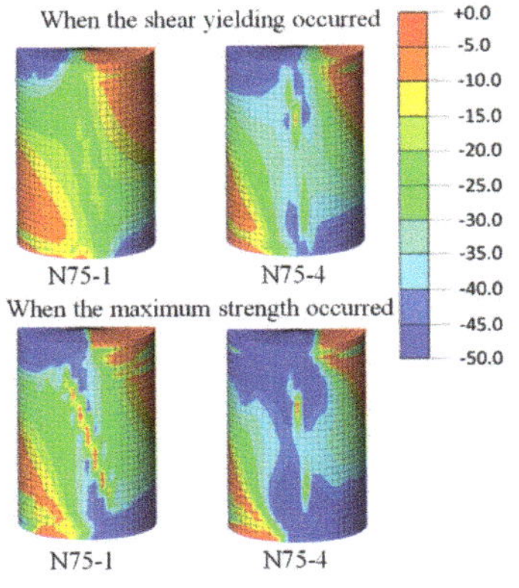

Figure 15. Compressive stress contour of concrete.

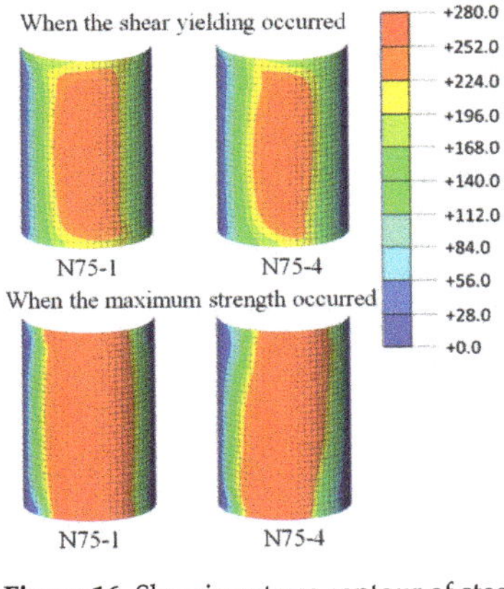

Figure 16. Shearing stress contour of steel tube.

6. Comparisons of Test, Analysis, and Calculation

In Figure 17 and Table 5, the maximum strength of the test $Q'max$, the maximum strength of the analysis *Qana*, shearing capacity *Qsu*, and flexural capacity *Qbu* are compared. The values of $Q'max$ are averages of the absolute values of *Qmax* and *Qmin* of the tests. The *Qsu* and *Qbu* are calculated using the methods defined in the CFST Recommendations in Japan. The *Qsu* is the generalized superposed strength, as shown in Equation (2).

$$Q_{su} = {}_sQ_{su} + {}_cQ_{su} \tag{2}$$

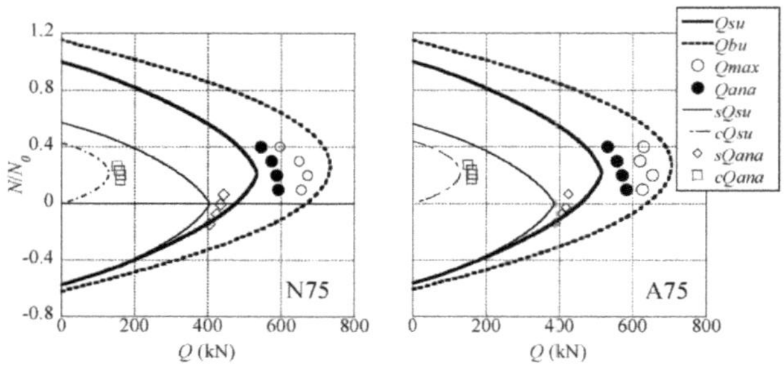

Figure 17. Relations between axial force ratio and shearing capacity.

Table 5. Comparisons of maximum strengths.

Specimen	$Q'max$ (kN)	*Qana* (kN)	*Qsu* (kN)	*Qbu* (kN)	$Q'max/Qana$	$Q'max/Qsu$	*Qana/Qsu*	*Qana/Qbu*
N75-1	655	592	506	703	1.11	1.29	1.17	0.84
N75-2	672	588	517	725	1.14	1.30	1.14	0.81
N75-3	649	573	511	728	1.13	1.27	1.12	0.79
N75-4	596	545	489	712	1.09	1.22	1.12	0.77
A75-1	627	582	486	671	1.08	1.29	1.20	0.87
A75-2	654	571	496	694	1.15	1.32	1.15	0.82
A75-3	620	557	490	698	1.11	1.27	1.14	0.80
A75-4	629	531	468	683	1.19	1.34	1.13	0.78
		Ave.			1.12	1.29	1.15	0.81
		St. dev.			0.033	0.034	0.025	0.031

The shearing capacities of steel tube and inner concrete are calculated by CFST Recommendations and are shown as *sQsu* and *cQsu*. The experimental results of $Q'max$ are greater than the analytical results of *Qana* and the calculated values of *Qsu* in all specimens. Because the average of the $Q'max/Qana$ is 1.12, the analyses can safely evaluate the maximum strengths of the experiments in 12%. The average values of $Qmax/Qsu$ and $Qana/Qsu$ are 1.29 and 1.15. The standard deviations of each ratio are also shown in Table 5. This shows that the methods in the CFST Recommendations provide safe estimates of the shearing capacities. Both $Q'max$ and *Qana* are lower than *Qbu*. This shows that the maximum strengths of the tests and analyses do not attain the full plastic states of the sections of the specimens, which are defined as the flexural capacities in the CFST Recommendations. The maximum strengths of the specimens are significantly affected by the shearing stress on the steel tube and inner concrete based on the observations of the tests and analyses. The comparisons between the test results and the design formulas suggest that all specimens fail in shear and exhibit their shearing capacities as the maximum strengths.

In Figure 17 and Table 6, the shear forces that the steel tube and inner concrete sustain at the maximum strength of the CFST are shown as *sQana* and *cQana* in the analyses and are compared with each other. *sQana* and *cQana* are estimated by integrating the bearing forces of the spring elements arranged in the middle surface of the specimen. The ratio of the shearing forces of the steel tube and inner concrete is approximately 7:3 for the specimens.

sQana/*sQsu* is 1.10 on average, so *sQsu* safely evaluates the shearing strength of the steel tube. The average value of *cQana*/*cQsu* is 1.29; thus, *cQsu* underestimates the shearing strength of the inner concrete.

Table 6. Comparisons between *sQana*, *cQana*, and *sQsu*, *cQsu*.

Specimen	*sQana* (kN)	*cQana* (kN)	*sQana*/*Qana*	*cQana*/*Qana*	*sQana*/*sQsu*	*cQana*/*cQsu*
N75-1	442	161	0.73	0.27	1.14	1.30
N75-2	434	163			1.08	1.27
N75-3	424	159			1.08	1.26
N75-4	404	151			1.09	1.34
A75-1	424	161	0.72	0.28	1.15	1.32
A75-2	417	164			1.09	1.27
A75-3	405	161			1.08	1.28
A75-4	389	150			1.09	1.32
	Ave.		0.72	0.28	1.10	1.29

7. Conclusions

Shearing capacities and behaviors of short CFST columns have been experimentally and analytically investigated. In the experiments, eight specimens with a/D of 0.75 are subjected to cyclic shearing forces under a constant axial force. The main test parameter was the axial force ratio. Relations between the shearing force and drift angle were obtained through the tests. Then, a 3D-FEM analysis was carried out to trace the load versus deformation relations of each specimen of the tests. The maximum strengths of tests and analyses were compared to each other and to the shearing and flexural capacities calculated by the design formulas of "Recommendations for Design of and Construction of Concrete Filled Steel Tubular Structures" in Japan. The following conclusions were derived.

1. In the tests, shear yielding of the steel tube was observed before each specimen exhibited the maximum strength. Diagonal cracks were observed in the inner concrete after the tests were conducted.
2. The results of the 3D-FEM analyses traced the overall trend of the shearing force and drift angle relations obtained from the tests.
3. The maximum strengths of the specimens obtained from the test and analysis were lower than the flexural capacities calculated by the CFST Recommendations in Japan. The load-carrying capacities of the CFST short columns were influenced by shearing forces.
4. The maximum strengths of the tests were compared with the shearing capacities calculated by the CFST Recommendations in Japan. The maximum strengths of the tests were underestimated, with 29% errors. The value of the standard deviation of the ratio of the test and calculation was 0.034.
5. The maximum strength of the tests was the evaluated safety, with a 12% error by the 3D-FEM analyses. The value of the standard deviation of the ratio of the test and analysis was 0.033.

Author Contributions: H.N. was responsible for study design. H.N., K.U. and Y.Y. were involved in experimental works, calculations, and data interpretation. All authors critically revised the report, commented on drafts of the manuscript, and approved the final report. All authors have read and agreed to the published version of the manuscript.

Funding: This research was founded by the Ministry of Education of Japan, Culture, Science, and Technology of Japan, under Grant-in-Aid for Young Scientists (B, No. 23760528, Representative: Hiroyuki Nakahara).

Data Availability Statement: The data that support the findings of this study are available from the corresponding author upon reasonable request.

Acknowledgments: The authors wish to express their thanks to H. Kubotera, technician of Kyushu University, and K. Kin and R. Tsumura, graduate students of Kyushu University, for their assistance in the experiment and preparation of this paper.

Conflicts of Interest: Author Yuto Yanai was employed by the company General Building Research Corporation of Japan. The remaining authors declare that the research was conducted in the absence of any commercial or financial relationships that could be construed as a potential conflict of interest.

References

1. El-Tawil, S.; Bracci, J. Recent Findings from Phase V of the United States-Japan Cooperative Earthquake Research Program. *ASCE J. Struct. Eng.* **2004**, *130*, 155–156. [CrossRef]
2. Goel, S.C. United States-Japan Cooperative Earthquake Engineering Research Program on Composite and Hybrid Structures. *ASCE J. Struct. Eng.* **2004**, *130*, 157–158. [CrossRef]
3. Spacone, E.; El-Tawil, S. Nonlinear Analysis of Steel-Concrete Composite Structures: State of the Art. *ASCE J. Struct. Eng.* **2004**, *130*, 159–168. [CrossRef]
4. Varma, A.H.; Ricles, J.M.; Sause, R.; Lu, L.W. Seismic Behavior and Design of High-Strength Square Concrete-Filled Steel Tube Beam Columns. *ASCE J. Struct. Eng.* **2004**, *130*, 169–179. [CrossRef]
5. Sakino, K.; Nakahara, H.; Morino, S.; Nishiyama, I. Behavior of Centrally-Loaded Concrete-Filled Steel Tube Short Columns. *ASCE J. Struct. Eng.* **2004**, *130*, 180–188. [CrossRef]
6. Inai, E.; Mukai, A.; Kai, M.; Tokinoya, H.; Fukumoto, T.; Koji Morita, K. Behavior of Concrete-Filled Steel Tube Beam Columns. *ASCE J. Struct. Eng.* **2004**, *130*, 189–202. [CrossRef]
7. Fujimoto, T.; Mukai, A.; Nishiyama, I.; Sakino, K. Behavior of Eccentrically Loaded Concrete-Filled Steel Tubular Columns. *ASCE J. Struct. Eng.* **2004**, *130*, 203–212. [CrossRef]
8. Azizinamini, A.; Schneider, S.P. Moment Connections to Circular Concrete-Filled Steel Tube Columns. *ASCE J. Struct. Eng.* **2004**, *130*, 213–222. [CrossRef]
9. Ricles, J.M.; Peng, S.W.; Lu, L.W. Seismic Behavior of Composite Concrete Filled Steel Tube Column-Wide Flange Beam Moment Connections. *ASCE J. Struct. Eng.* **2004**, *130*, 223–232. [CrossRef]
10. MacRae, G.; Roeder, C.W.; Gunderson, C.; Kimura, Y. Brace-Beam-Column Connections for Concentrically Braced Frames with Concrete Filled Tube Columns. *ASCE J. Struct. Eng.* **2004**, *130*, 233–243. [CrossRef]
11. Nishiyama, I.; Fujimoto, T.; Fukumoto, T.; Yoshioka, K. Inelastic Force-Deformation Response of Joint Shear Panels in Beam-Column Moment Connections to Concrete-Filled Tubes. *ASCE J. Struct. Eng.* **2004**, *130*, 244–252. [CrossRef]
12. Nakahara, H.; Sakino, K. Hysteretic Behavior of Concrete Filled Steel Tubular Columns under Uniform Bending. *ACI Int. Symp. Confin. Concr.* **2006**, *238*, 289–304.
13. Koester, B.D.; Uchida, K.; Noguchi, H.; Yura, J.A.; Jirsa, J.O. *Panel-Zone Behavior of Moment Connections Between Steel Beams and Concrete-Filled Steel Tube Columns, Proceedings of the Structural Engineers World Congress, San Francisco, CA, USA, 18–23 July 1998*; Elsevier Science: Amsterdam, The Netherlands, 1998.
14. Noguchi, H.; Uchida, K. Finite Element Method Analysis of Hybrid Structural Frames with Reinforced Concrete Columns and Steel Beams. *ASCE J. Struct. Eng.* **2004**, *130*, 328–335. [CrossRef]
15. Architectural Institute of Japan (AIJ). *Recommendations for Design and Construction of Concrete Filled Steel Tubular Structures*; Architectural Institute of Tokyo: Tokyo, Japan, 2008.
16. Sakino, K.; Ishibashi, H. Experimental Studies on Concrete Filled Square Steel Tubular Short Columns Subjected to Cyclic Shearing Force and Constant Axial Force. *AIJ J. Struct. Constr. Eng.* **1985**, *353*, 81–91. [CrossRef]
17. Choi, K.K.; Xiao, Y. Analytical Studies of Concrete-Filled Circular Steel Tubes under Axial Compression. *ASCE J. Struct. Eng.* **2010**, *136*, 565–573. [CrossRef]
18. Hu, H.T.; Huang, C.S.; Wu, M.H.; Wu, Y.M. Nonlinear Analysis of Axially Loaded Concrete-Filled Tube Columns with Confinement Effect. *ASCE J. Struct. Eng.* **2003**, *129*, 1322–1329. [CrossRef]
19. Varma, A.H.; Sause, R.; Ricles, J.M.; Li, Q. Development and Validation of Fiber Model for High-Strength Square Concrete-Filled Steel Tube Beam-Columns. *ACI Struct. J.* **2005**, *102*, 73–85.
20. Chiew, S.P.; Lie, S.T.; Dai, C.W. Moment Resistance of Steel I-Beam to CFST Column Connections. *ASCE J. Struct. Eng.* **2001**, *127*, 1164–1172. [CrossRef]
21. Tort, C.; Hajjar, J.F. Mixed Finite-Element Modelling of Rectangular Concrete-Filled Steel Tube Members and Frames under Static and Dynamic Loads. *ASCE J. Struct. Eng.* **2010**, *136*, 654–664. [CrossRef]
22. Roeder, C.; Lehman, D.; Heid, A.; Maki, T. Shear Design Expressions for Concrete Filled Steel Tube and Reinforced Concrete Filled Tube Components. WSDOT Research Report. Seattle, Washington. 2016. Available online: https://www.wsdot.wa.gov/research/reports/fullreports/776.2.pdf (accessed on 12 June 2016).
23. Vecchio, F.J.; Collins, M.P. The Modified compression-Field Theory for Reinforced Concrete Elements Subjected to Shear. *ACI J.* **1986**, *83*, 219–231.
24. Architectural Institute of Japan (AIJ). *Design Guidelines for Earthquake Resistant Reinforced Concrete Building Based on Inelastic Displacement Concept*; Architectural Institute of Japan: Tokyo, Japan, 2010.
25. Hordjik, D.A. Local Approach to Fatigue of Concrete. Ph.D. Thesis, Delft University of Technology, Delft, The Netherlands, 1991.
26. Japan Society of Civil Engineers (JSCE). *Standard Specifications for Concrete Structures*; Japan Society of Civil Engineers: Tokyo, Japan, 2012.
27. Feenstra, P.H. Computational Aspects of Biaxial Stress in Plain and Reinforced Concrete. Ph.D. Thesis, Delft University of Technology, Delft, The Netherlands, 1993.

28. Nakamura, H.; Higai, T. Compressive Fracture Energy and Fracture Zone Length of Concrete. In *JCI-C51E Seminar on Post-Peak Behavior of RC Structures Subjected to Seismic Loads*; Japan Concrete Institute: Tokyo, Japan, 2001; pp. 259–272.
29. Komuro, M.; Kishi, N.; Zhang, G. Analytical Study on Load-Carrying of Partially Concrete-Filled Steel-Pipe Pier Models. *JSCE J. Appl. Mech.* **2003**, *6*, 475–486. [CrossRef]

buildings

MDPI

Article

Effects of Out-of-Plane Deformation of the Base Plate on the Structural Behavior of an Exposed Column Base

Ding Nan, Hiroyuki Nakahara * and Iathong Chan

Graduate School of Engineering, Nagasaki University, Nagasaki 852-8521, Japan; bb52222101@ms.nagasaki-u.ac.jp (D.N.); chan.iathong@nagasaki-u.ac.jp (I.C.)
* Correspondence: nakaharahiroyuki@nagasaki-u.ac.jp; Tel.: +81-819-2895

Abstract: This study explores the behavior of exposed column bases in concrete-filled steel tubular (CFST) and steel structures, with a focus on cases where base plates yield due to out-of-plane deformation. Understanding these mechanisms is crucial for improving the design and safety of these structures. Experimental tests and numerical analyses were conducted on four specimens to investigate their lateral load versus drift angle behavior. The tests demonstrated that base plates exhibited sufficient deformation capacities and enhanced hysteresis characteristics. Finite element method (FEM) analysis successfully traced the load–deformation relationships observed in the tests, providing detailed insights into stress distribution on the base plates. Based on these analyses, a simplified calculation method was proposed to evaluate the horizontal strength of exposed column bases. The proposed method showed excellent agreement with the test results, highlighting its potential as a practical tool for structural design.

Keywords: exposed column base; elastic plastic behavior; FEM analysis; hinging line theory

1. Introduction

Concrete-filled steel tubular (CFST) structures have been used in seismic regions. A CFST structure consists of CFST columns, steel beams, and steel base plates. This structure is categorized in steel structure. There have been many steel-structure buildings that collapsed due to damage to exposed column bases, for example, as shown in Figure 1. These damages were reported in [1–3].

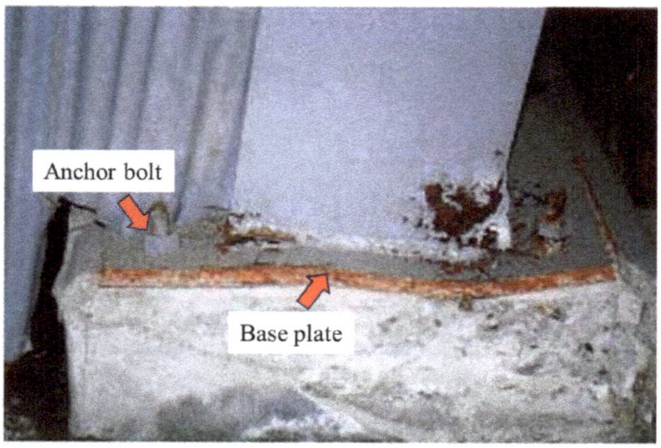

Figure 1. Damage of an exposed column base (reprinted from [1]).

An exposed column base connection is commonly used in mid- and low-rise buildings owing to its ease of installation and cost effectiveness. Several studies have shown interesting outcomes for column bases [4–7]. The base plate is welded to the lower end of the column and is connected by anchor bolts to the concrete foundation. The authors focus on the structural behavior and the calculation method of the strength of an exposed column base which yields at the base plate.

Figure 2 shows the statistical data of the column bases of steel structures damaged by the Kobe earthquake in 1995 in Japan [1]. Anchor bolt failures account for 87% of the data. This is because the current design guidelines [8] in Japan specify that anchor bolts in exposed column bases should yield axially first, preventing the base plate from plasticizing. Anchor bolt failures were studied in some articles [9–13]. However, column bases with fracturing anchor bolts have a poor energy absorption capacity [14]. Past earthquake damage has shown not only insufficient elongation of anchor bolts but also significant out-of-plane deformations of base plates.

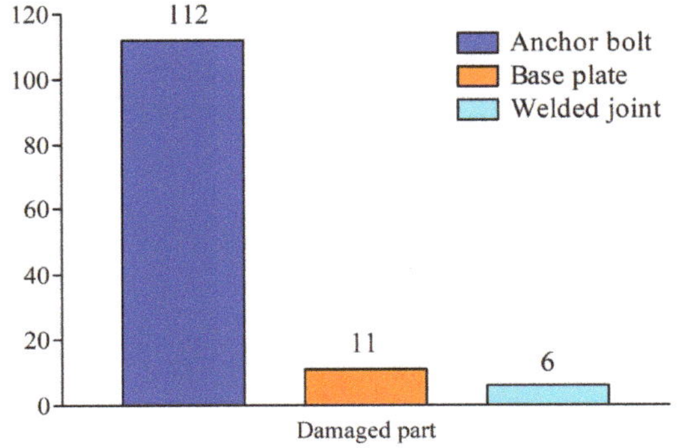

Figure 2. Number of parts in exposed column bases in steel structures damaged by the Kobe earthquake (data from [1]).

Studies of the structural design including out-of-plane deformation of the base plate in an exposed column base were reported in [15,16].

In a study by Kokubo et al. [17], the yielding type of anchor bolts and base plate were tested. In the test, cyclic loads were applied in the horizontal direction with a displacement control under a constant axial force of 200 kN. The history of the drift angles of the column was ±0.5%, ±1%, ±2%, ±4%, and ±8%. The study confirmed that the hysteresis characteristics of the base plate yielding type were not inferior to those of the anchor bolts yielding type.

Furthermore, in an experimental study by Akiyama et al. [18], the test parameter was thickness of the base plate of 60 mm and 30 mm. The former was an anchor bolt yielding type column base and the latter was a base plate yielding type column base. An inertial loading device was used to vibrate the test specimen, and vibrations were applied in the order of elastic excitation, plastic excitation, and ultimate excitation. In both cases, the maximum rotation angles of the column bases were around 6/100, demonstrating large deformation performance. In the base plate yielding type, the base plate deformed out of plane, and the hysteresis characteristics were intermediate between those of the slip type and the origin-oriented type.

In a study by Takamatsu et al. [19], a total of four specimens with base plates of 9 mm and 19 mm thickness were tested under axial forces of 0 kN and 500 kN. A predetermined axial force was applied using a vertical hydraulic jack, and cyclic horizontal loadings were conducted using a horizontal hydraulic jack. The horizontal loading was repeated in a displacement-controlled manner for each cycle until the rotational angle at the column base reached approximately 0.1 rad. The test results showed that for the specimens without axial force, the base plate plastically deformed out of plane, and the hysteresis curve exhibited typical origin-oriented characteristics. The hysteresis curve for the specimens with an axial force of 500 kN also showed origin-oriented characteristics.

Based on the above studies, out-of-plane deformation of the base plate in exposed column bases is not inevitable to avoid as a failure mode in the critical state of buildings in large-scale earthquakes. From the point of view of deformation capacities and energy-dissipating capacities of the parts, it is necessary to accumulate the test and analytical studies of the base plate yielding type of the column base.

Exposed column bases are also used in CFST (concrete-filled steel tubular) structures. Due to the high strength of CFST structures, it is difficult to realize that column yielding mechanism before the yielding of the part of the column base.

The authors thus conducted a lateral loading test of four specimens of column bases with CFST columns and derived the strength formula for column bases with yielding at the base plates by comparing the test and FEM analysis results carried out to clarify the load versus deformation relationship and stress distribution states of the base plates.

2. Test Study

To demonstrate the behavior of exposed column bases where the base plate yields first, tests were conducted as shown in Figure 3. The specimen consisted of a CFST column, a thin steel base plate, and a rigid steel foundation.

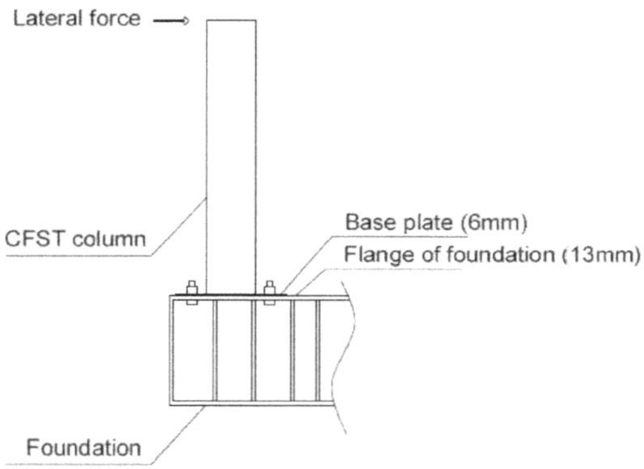

Figure 3. Test specimen.

2.1. Specimens

The dimensions of the base plate are shown in Figure 4. The test specimens were made of □-150 × 150 × 4.50 mm STKR400 square steel tubes. The height of all specimens was 855 mm. The material properties of the column are shown in Table 1. The steel plates for the base plates measured 220 × 335 mm, with thicknesses of 6.00 mm and 5.60 mm. The material properties of the plates are shown in Table 2. The bolt holes were 22 mm in diameter. The steel plates were welded to the steel tubes. The specimens were fixed to the loading frame using four high-strength M20 bolts. Strain gauges were glued on the base plates to measure the deformations at each place. Careful measurement of the strain on the surface of the plates was demonstrated. In specimens S1 and S2, the positions of strain gauges were P-1, P-3, P-4, and P-6. In specimens S3 and S4, the positions of strain gauges were P-2, P-3, P-4, and P-5. In this study, the columns are not steel but CFST, in which concrete filled steel tubes. The compressive strengths of the concrete are shown in Table 3. The notations in the tables are as follows: σy: yield stress; σu: tensile stress; σc: compressive strength of concrete; E: Young's modulus; Ec: Young's modulus of the concrete; B_P: width of the base plate; D_P: length of the base plate; t: thickness of the base plate.

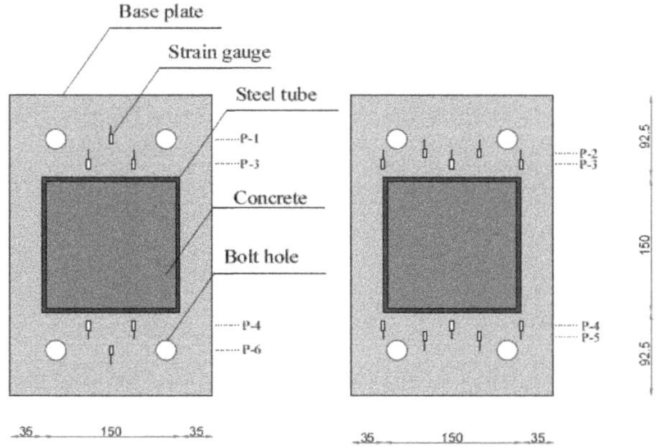

Figure 4. Base plates and locations of strain gauges (in mm).

Table 1. Material properties of the columns.

Specimen	σ_y (N/mm^2)	σ_u (N/mm^2)	E (kN/mm^2)
S1 S2	348	446	194
S3 S4	352	449	205

Table 2. Dimensions and material properties of the base plate.

Specimen	B_P (mm)	D_P (mm)	t (mm)	σy (N/mm^2)	σu (N/mm^2)	E (kN/mm^2)
S1 S2	220	335	6.00	344	440	205
S3 S4	220	335	5.60	294	454	199

Table 3. Material properties of concrete.

Specimen	σc (N/mm^2)	Ec (kN/mm^2)
S1 S2	47.6	31.7
S3 S4	49.5	30.2

2.2. Test Setup

The tests were conducted with two setups, as shown in Figures 5 and 6. The foundation of the specimen was not concrete but steel because the test focused on out-of-plane deformation of the base plate by using a rigid steel foundation.

The equipment for setup 1 consisted of a 100 kN capacity hydraulic jack, an SS400 steel frame, and a 200 kN capacity load cell. An axial force was applied by introducing tensile forces to two PC steel rods set longitudinally. Tensile strain was measured with strain gauges glued on the PC steel rods. The equipment for setup 2 consisted of a 300 kN capacity double-acting hydraulic jack. An axial force was not applied. All the specimens were set to rigid foundations with four M20 high-strength bolts.

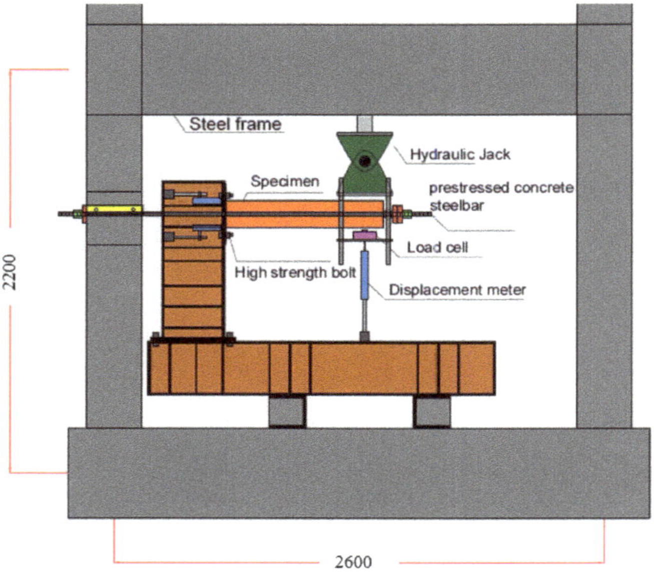

Figure 5. Loading apparatus (setup 1).

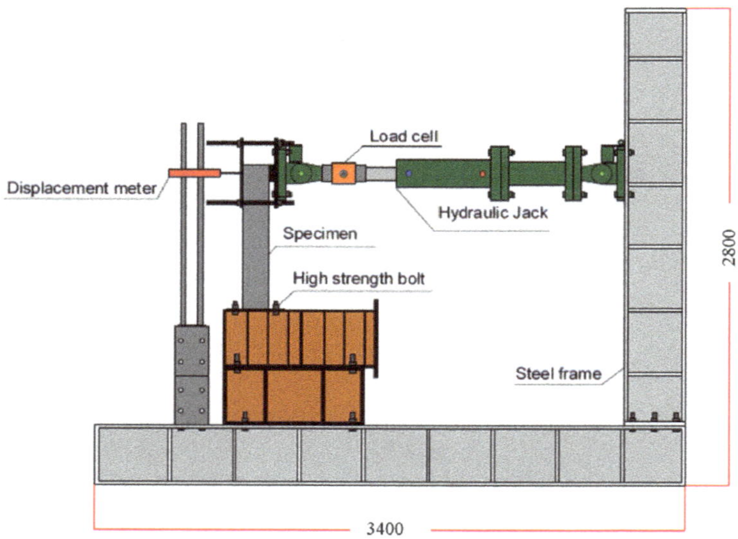

Figure 6. Loading apparatus (setup 2).

The axial force of the S1 and S2 specimens was introduced from the top of the column by tightening the PC steel rods. The value of the axial force was 50 kN. In setup 1, the lateral force was applied at a height of 760 mm, and at a height of 810 mm in setup 2. Cyclic loadings were carried out under displacement control obtained with an LVST (linear variable differential transformer) at the loading point. The drift angles of the columns of 0.01, 0.02, 0.03, and 0.05 rad were applied twice in each step.

2.3. Test Results and Discussion

The test results of lateral force versus drift angle are shown in Figure 7, where the restoring force characteristics are stable even at the drift angle of 0.05 rad. The energy absorption capacities were maintained when the base plates were thin, 6 mm and 5.6 mm. After the loading tests, plastic deformations were observed only in the base plates, with no residual deformations in the bolts or the columns.

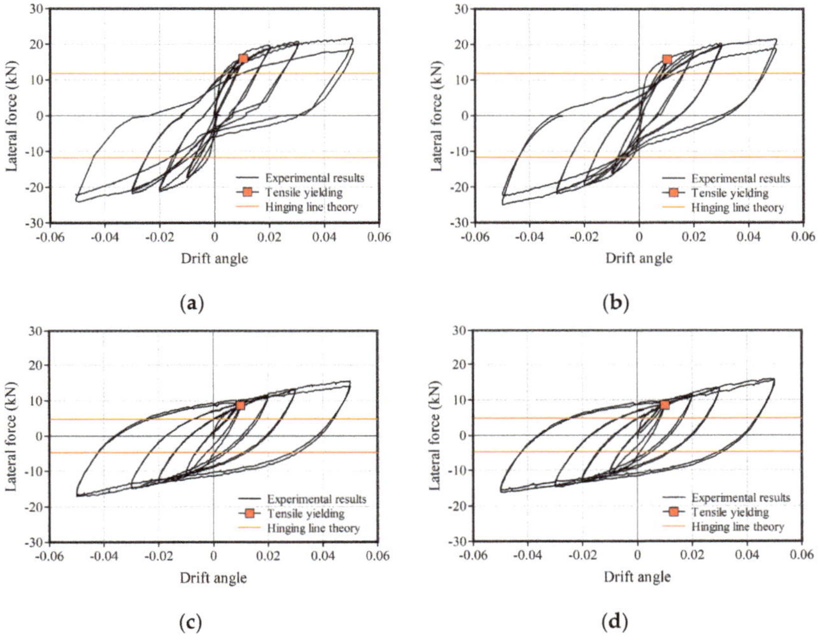

Figure 7. Test results: (**a**) S1, (**b**) S2, (**c**) S3, and (**d**) S4.

The lateral force corresponding to the yield strength and the full plastic strength of the steel tube of S1 and S2 were 56.4 kN and 84.7 kN, respectively. Those of the steel tubes of S3 and S4 were 43.6 kN and 65.4 kN, respectively. These values were sufficiently larger than the yielding strength of the base plates.

There are two declining points in the load–deformation curves in Figure 7. Stiffness obviously declined at drift angle R of around 0.002 rad and 0.01 rad. The first was due to out-of-plane deformation of the base plate and the second was due to the tensile yielding of the base plate.

The strength of the base plates calculated using the theory of the yield hinging lines was 11.8 kN for the S1 and S2 specimens and 4.81 kN for the S3 and S4 specimens. These are shown as orange lines in Figure 7. At the early stage of the loading, the base plate started plastic bending at around R = 0.002 rad, the strength of which was estimated using the hinging line method. The hinging line theory tends to underestimate the second yielding points of the test results. The authors proposed the calculating method to estimate the second yielding strength of the test accurately.

2.4. Discussion of the Deformation of the Base Plates

One example of strain at the local point of P-4 on the base plate of specimen S1 is shown in Figure 8. Strain elongation was greater than the shrinkage. This indicates that the strain energy of the base plate under the tensile force must be considered to estimate the strength. The authors focused on the tensile yielding points of the base plates to evaluate the strength of exposed column bases. The yield point of all the specimens was determined at the drift angle of 0.01 rad as explained later.

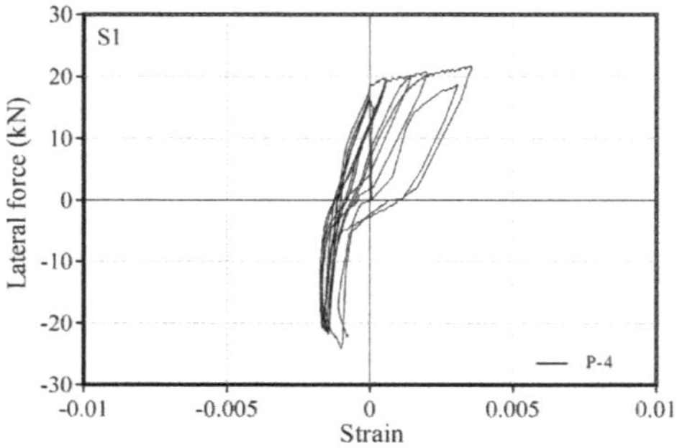

Figure 8. Relationship between lateral force and local strain at P-4 of S1.

2.5. Study of the Equivalent Viscous Damping Ratio

The value of the equivalent viscous damping ratio (h_{eq}) based on the load–deformation relationship of the test is useful to estimate the energy absorption capacity. Figure 9 shows the hysteresis curve of the S1 specimen in the first cycle. The examination of the equivalent viscous damping ratio for each cycle is presented in Table 4. It can be observed that the value of h_{eq} increased with more cycles. This means that the damping effect of the column base increased with the increase in plastic deformation. The average value of the equivalent viscous damping ratio was 20.7%, which is higher than the value of the damping ratio of 2% in a dynamic response analysis in normal seismic designs.

$$h_{eq} = \frac{1}{4\pi}\left(\frac{\Delta W}{W_e}\right) \quad (1)$$

ΔW : Area of one cycle of the hysteresis loop
W_e : Elastic srain energy

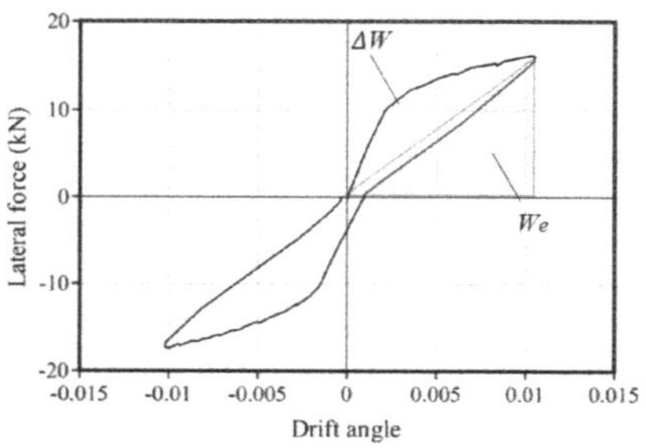

Figure 9. Hysteresis loop of the test and strain energy.

Table 4. Equivalent viscous damping ratio for each cycle.

Specimen	h_{eq} (%)				Average h_{eq} (%)		
	0.01 rad	0.02 rad	0.03 rad	0.05 rad			
S1	7.68	13.9	16.4	23.5	15.4	16.0	20.7
S2	7.78	14.7	18.8	25.2	16.6		
S3	16.4	23.3	28.7	35.2	25.9	25.5	
S4	15.3	23.5	28.3	33.2	25.1		

3. Analysis of the Base Plate

Three-dimensional finite element method (FEM) analysis was performed using software Marc 2023 SE.

3.1. Model and Loading Process

The FEM analysis model is shown in Figure 10. The specimen was touched on the infinite rigid body of the foundation. The base plate was fixed to the rigid body at the points of bolt holes. The axial forces of S1 and S2 were subjected to 50 kN in the *Z*-axis direction. The lateral force was applied in the *y*-axis direction under the control of the displacement conducted in the test. The stress–strain relationships shown in Figure 11 were used in the analysis. The values of Young's modulus and yield strength of the steels were the same as those in Table 2. Poisson's ratio was 0.3 for the analysis. The von Mises yield criterion shown in Figure 12 was used under the triaxial stress state of the steels.

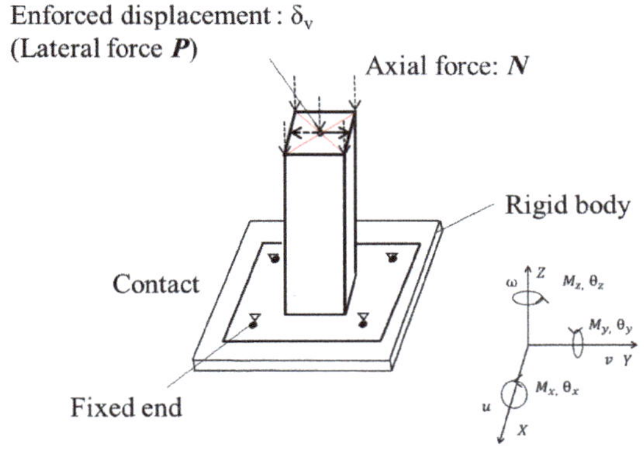

Figure 10. Boundary condition and loading direction for the FEM analysis.

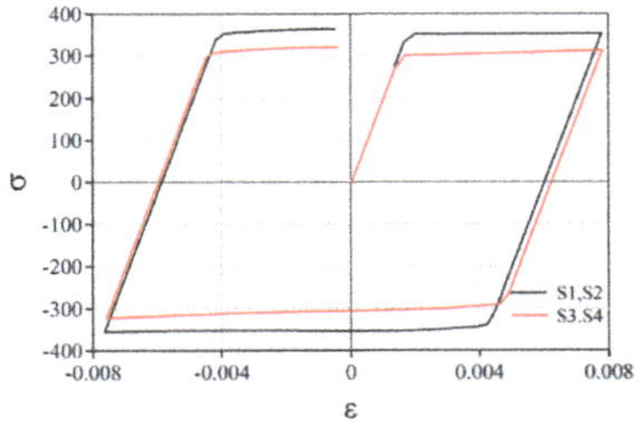

Figure 11. Stress–strain relationship.

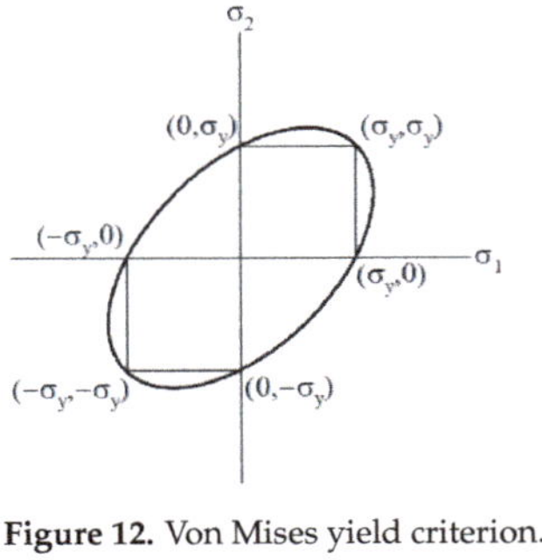

Figure 12. Von Mises yield criterion.

3.2. Analytical Results and Discussion

The results of the analysis (red lines) are shown in Figure 13, compared to the test results (black lines) within the absolute value of the drift angle of 0.01 rad. The analytical results traced well to the test results.

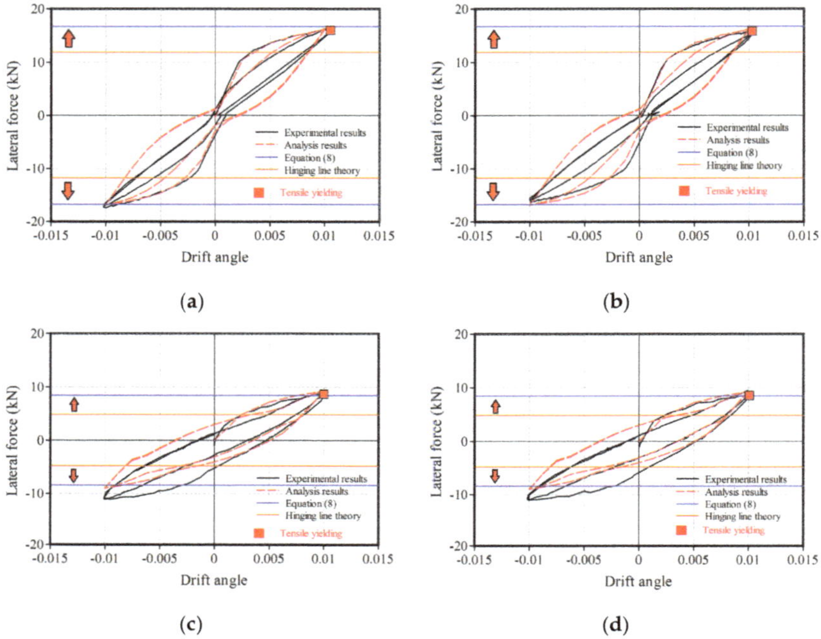

Figure 13. Analysis results: (**a**) S1; (**b**) S2; (**c**) S3; (**d**) S4.

In the load–deformation curves of the test and the analysis, the stiffness obviously declined at the drift angle R of around 0.002 rad and 0.01 rad. At the early stage of loading, the base plate started to bend at around R = 0.002 rad, the strength of which was estimated using the hinging line method. After R = 0.01 rad, the strength would not increase from the test result in Figure 7. The authors focused on the strength at R = 0.01 rad since that strength is necessary for a rational design of the plates of an exposed column base. Figures 14 and 15 show von Mises stress contours and Y-directional stress contours for the S1 and S2 analyses at R = 0.01 rad. Figures 16 and 17 show von Mises stress contours and Y-directional stress contours for the S3 and S4 analyses at R = 0.01 rad. In this analysis, the Y-direction was the same as the loading direction of the test. The warm-colored area was found to be the area of stress concentration and yielding. In the Y-direction contours, the tensile side of the plate increased its stress. These figures show the phenomenon that the tensile resistance of the base plate increases with the progress of the out-of-plane deformation of the base plate. It was found that tensile yielding at the base plate in the test and the analysis occurred at around analyzer yields at R = 0.01 rad.

In the study, the authors proposed a calculation method which is shown with the blue lines in Figure 13. The red arrows indicate that the values obtained using the strength formula proposed by the authors exceed the calculated values based on the hinging line method. The strains at the position of P-4 on the base plate obtained in the test and the analysis attained its yield strain ε_y in Figure 18, where it is evident that the strength of the column base must be estimated by considering the tensile resistance of the base plate. A detailed explanation of the calculation method is provided in the next section.

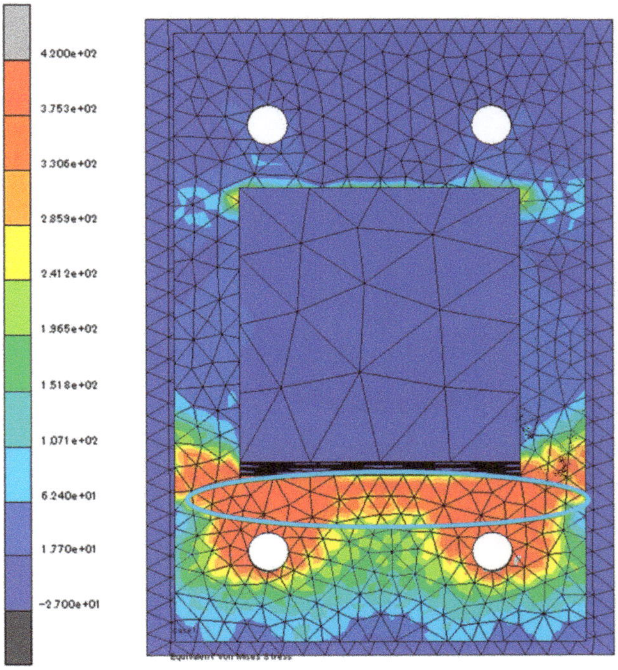

Figure 14. Analytical von Mises stress contours for S1 and S2 (at R = 0.01 rad).

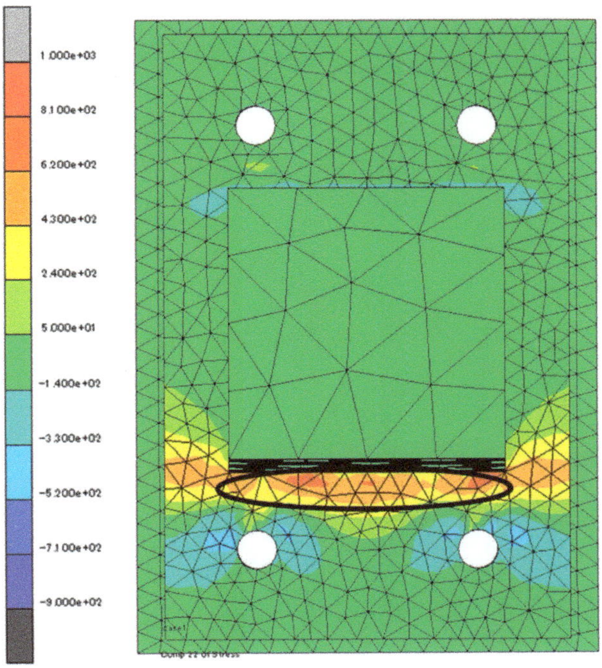

Figure 15. Analytical Y-directional stress contours for S1 and S2 (at R = 0.01 rad).

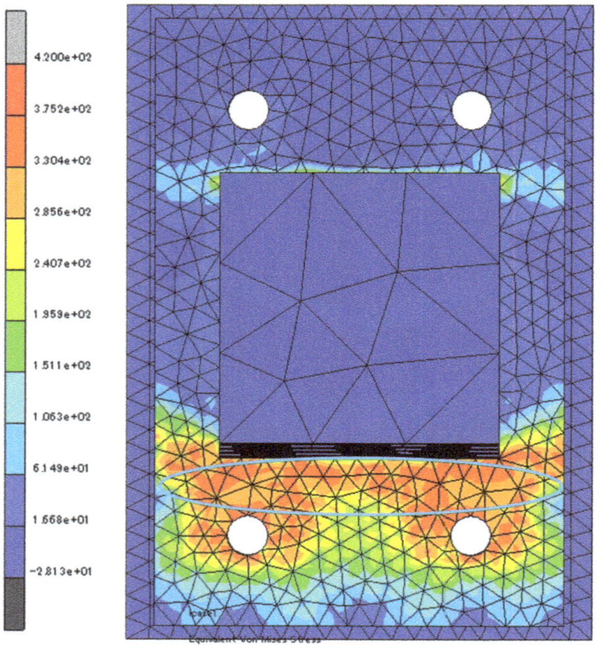

Figure 16. Analytical Mises stress contours for S3 and S4 (at R = 0.01 rad).

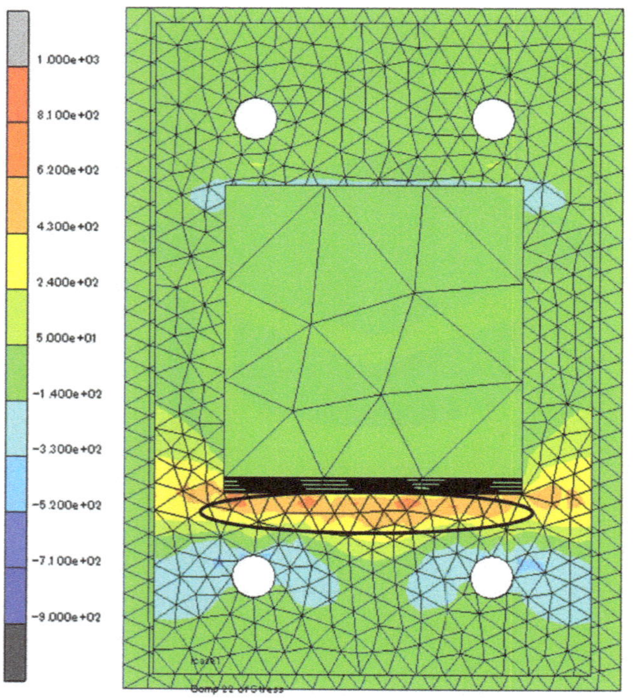

Figure 17. Analytical Y-directional stress contours for S3 and S4 (at R = 0.01 rad).

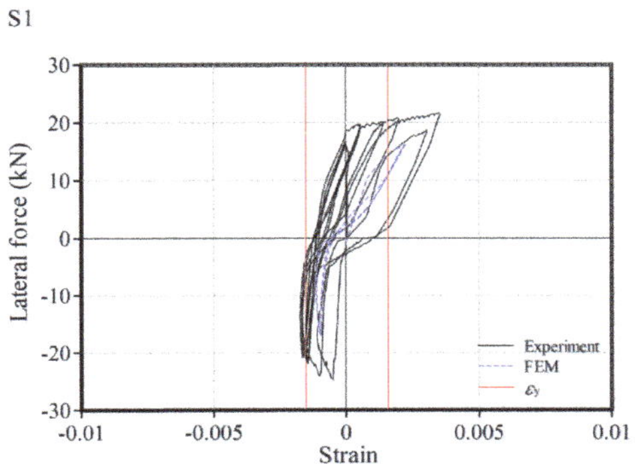

Figure 18. Comparison of the strain at the plate versus lateral force between the test and the analysis.

4. Simple Calculation Method for the Yielding Strength of an Exposed Column Base

Rational structural designs need to predict the yielding strength of parts precisely. A simple calculation method is proposed to evaluate the yielding strength of the base plates through the hinging line theory [20] and strain energy theory.

The failure mechanism is assumed to be illustrated in Figures 19 and 20 under a lateral force *P*. Figure 20 illustrates the deformation of the plate when the top of the column sways at the drift angle *R*, where the red points show the hinging lines and the blue part shows the elongation part. The internal work *U* includes the plastic rotating deformation at the hinging line and the elastic elongation deformation of the base plate, assuming that the column and bolts are rigid.

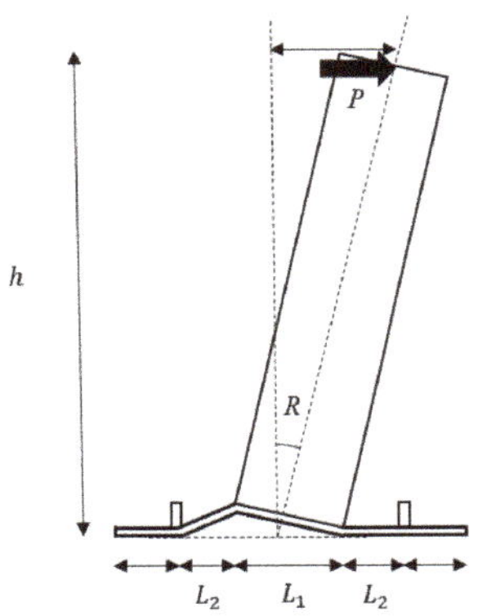

Figure 19. Deformation of the test specimen at drift angle *R*.

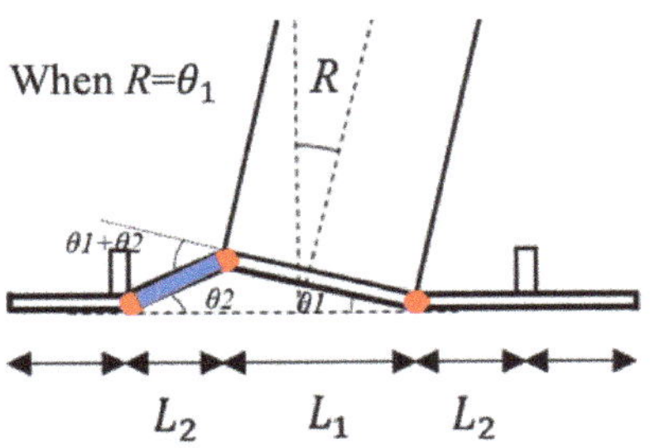

Figure 20. Local deformation of the base plate.

Figure 21 shows the plan view of three hinging lines and the elongation area. In the analysis, plastic out-of-plane deformations of the base plates occurred on the three lines at the drift angle of 0.05 rad. On the other hand, tensile strain distributions were getting wider in a trapezoidal shape on the base plates until $R = 0.01$ rad.

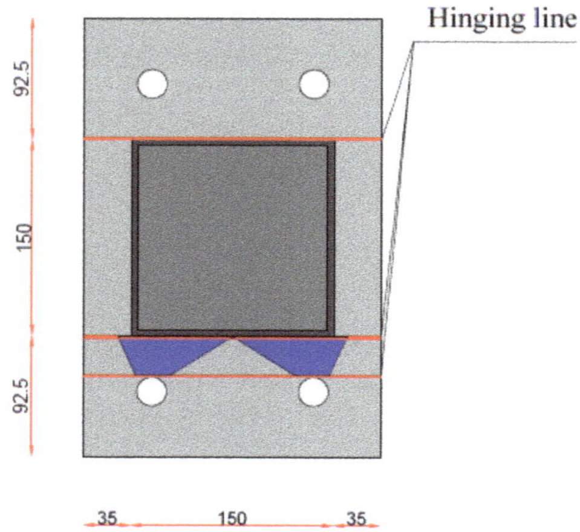

Figure 21. Hinging lines and tensile strain distribution areas on the base plate (plan view).

From Equation (2), the external work W due to the load was obtained. This is equal to the internal work U caused by deformation in the base plate shown in Equation (3). Equations (2) and (3) are expressed as the functions of the variable R. Equation (4) represents the formula for the full plastic moment of the plate. Equation (5) shows the geometric relationship between the tensile strain of the base plate and the drift angle.

$$W = P \cdot h \cdot R - N \cdot \frac{L_1}{2} \cdot R \tag{2}$$

$$U = 2M_p \alpha R + 2M_p R + \frac{EA{\varepsilon_{bp}}^2}{2} L_2 \tag{3}$$

$$M_p = \sigma_y \cdot B_p \cdot t^2/4, \tag{4}$$

$$\varepsilon_{bp} = \sqrt{1 + (\alpha R)^2} - 1 \tag{5}$$

The symbols used in the equations are summarized below:

A : Area of the tensile yielding part of the base plate
B_p : Width of the base plate
E : Young's modulus
h : Height at the loading point
L_1 : Depth of the steel tube
L_2 : Distance between the hinging lines
M_p : Full plastic moment of the base plate
N : Axial force
P : Lateral force
α : L_1/L_2
ε_{bp} : Tensile strain of the base plate

The first variation of the external work δW and the first variation of the internal work δU can be expressed by taking the variation of the variable R as follows:

$$\delta W = P \cdot h \cdot \delta R - N \cdot \frac{L_1}{2} \cdot \delta R \tag{6}$$

$$\delta U = 2M_p \alpha \delta R + 2M_p \delta R + \alpha^2 EAL_2 \left\{ R - \frac{R}{\sqrt{(\alpha R)^2 + 1}} \right\} \delta R \tag{7}$$

The principle of virtual work is expressed as $\delta W = \delta U$. This means Equation (6) = Equation (7), where the first variation of the drift angle δR is eliminated, so that:

$$P \cdot h = 2M_p \alpha + 2M_p + \alpha^2 EAL_2 \left\{ R - \frac{R}{\sqrt{(\alpha R)^2 + 1}} \right\} + \frac{NL_1}{2} \tag{8}$$

Figure 22 shows the relationship between the strain ε_{bp} on the base plate and the drift angle of the specimen. The observed strains except for the S2 specimen attained the yield strain ε_y of the base plate. The strain calculated with Equation (5) increased as the drift angle increased and exceeded the yield strain ε_y. From the equation, drift angle R is 0.0118nrad when $\varepsilon_{bp\cdot} = \varepsilon_y$. The strength of the column base was defined at R = 0.01 rad. Figure 23 shows the calculation models for the lateral force versus drift angle. The first declining point was defined at R = 0.002 rad, which was caused by the out-of-plane deformation of the base plate. The second declining point was defined at R = 0.01 rad, which was caused by the tensile yielding of the base plate. These are rough definitions. The authors will continue to develop a more precise model for the same concept in further studies.

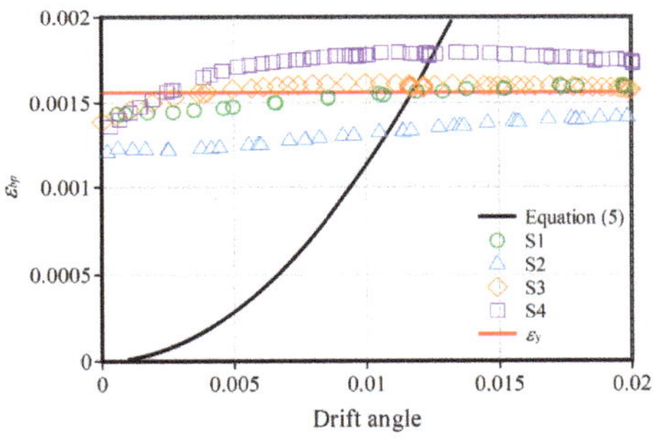

Figure 22. Relationship between the strain on the base plate and the drift angle.

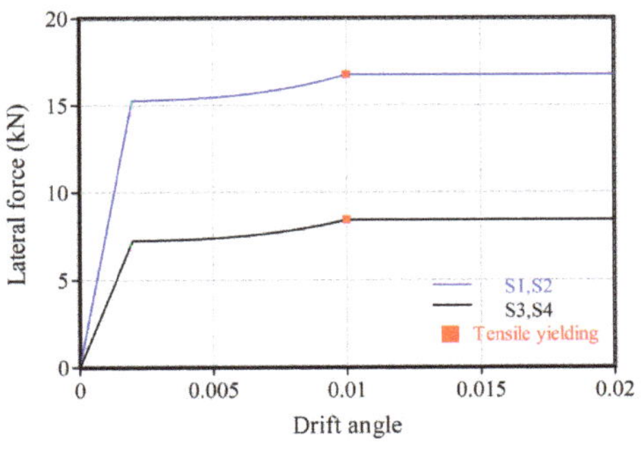

Figure 23. Calculation models for lateral force versus drift angle.

The calculated results for the tensile yielding strength of S1, S2 and S3, S4 were 16.7 kN and 8.41 kN, respectively. The test, calculated, and analytical values are summarized in

Table 5. The test results were predicted by calculations with 5.4% errors, while the analysis estimates resulted in 7.5% errors.

Table 5. Comparison of the test, calculated, and analytical values of yield strengths.

	Test (kN)	Calculation (kN)	Analysis (kN)	Exp./Cal. (%)	Exp./Ana. (%)
S1	16.0	16.7	16.8	95.8	95.2
S2	15.8		16.8	94.6	94.0
S3	8.69	8.41	9.23	103	94.1
S4	8.54		9.23	102	92.5

5. Conclusions

This study focused on the yield strength and yielding behavior of the base plate of an exposed column base using a test and the analysis. The simple calculation method was proposed on the basis of the strain energy theorem and the distribution of tensile stress in the base plate obtained by the FEM analysis. The findings are summarized below:

1. All the specimens failed in yielding at the base plates of an exposed CFST column base.
2. The test results of the relationship between the lateral load and the drift angle showed stability without strength deterioration. The hysteresis characteristics of the specimens with axial force application were origin-oriented, and those of the specimens with no axial force application were spindle-shaped.
3. Energy absorption capacities were estimated using the equivalent viscous damping ratio at each step of the drift angle of 0.01, 0.02, 0.03, and 0.05 rad. The ratios increased with cyclical loading. The average value was 20.7%.
4. The FEM analysis traced well the test results, revealing the tensile yielding area of the base plates.
5. The FEM analysis provided the stress distribution on the base plates when the load versus deformation relationship showed a significant decline at $R = 0.01$ rad.
6. The calculation method for the tensile yielding strength of the base plate of a column base accurately evaluated each test strength. It showed that the assumption of the stress distribution area was reasonable owing to the results of the FEM analysis.
7. The simple calculation method proposed in this paper demonstrated high consistency with the test results, suggesting its potential as a practical tool for structural design.

Author Contributions: H.N. was responsible for the study design. D.N., H.N., and I.C. were involved in the analysis, calculations, and data interpretation. All the authors critically revised the report, commented on the drafts of the manuscript, and approved the final report. All authors have read and agreed to the published version of the manuscript.

Funding: This study was supported by the following research: "Development and Practical Use of WGFT (Representative: Hiroyuki Nakahara at Nagasaki University)"; grant provided through the Construction Technology Research and Development Subsidy Program by the Ministry of Land, Infrastructure, Transport, and Tourism of Japan in 2023–2024.

Data Availability Statement: The data that support the findings of this study are available from the corresponding author upon reasonable request.

Acknowledgments: The authors wish to express their gratitude to Nanae Mikoshi, Ryo Aoki, and Shoichiro Sasaki (students of Nagasaki University) and Kenta Okamoto (technician at Nagasaki University) for their contributions to the experimental study.

Conflicts of Interest: The authors declare no conflict of interest.

References

1. *Reconnaissance Report on Damage to Steel Building Structures Observed from the 1995 Hyogoken-Nanbu Earthquake*; Steel Committee of Kinki Branch the Architectural Institute of Japan (AIJ): Osaka, Japan, 1995.

2. *Final Report of the Architectural Earthquake Investigation Committee on the 1995 Great Hanshin-Awaji Earthquake*; Architectural Earthquake Investigation Committee: Tokyo, Japan, 1996.
3. *Damage and Lessons of Steel Structures in the Hyogoken-Nanbu Earthquake*; Architectural Institute of Japan (AIJ): Tokyo, Japan, 1996.
4. Cui, Y.; Wang, F.; Yamada, S. Effect of Column Base Behavior on Seismic Performance of Multi-Story Steel Moment Resisting Frames. *Int. J. Struct. Stab. Dyn.* **2019**, *19*, 1940007. [CrossRef]
5. Cui, Y. Shear behavior of exposed column base connections. *Steel Compos. Struct.* **2016**, *21*, 357–371. [CrossRef]
6. Kishiki, S.; Yang, X.; Ishida, T.; Tatsumi, N.; Yamada, S. Experimental study of concrete breakout failure mechanism in an exposed column base with a foundation beam. *Eng. Struct.* **2021**, *243*, 112661. [CrossRef]
7. You, Y.-C.; Lee, D. Development of improved exposed column-base plate strong-axis joints of low-rise steel buildings. *J. Constr. Steel Res.* **2020**, *169*, 106062. [CrossRef]
8. *Recommendation for Design of Connection in Steel Structures*; Architectural Institute of Japan (AIJ): Tokyo, Japan, 2021.
9. You, Y.-C.; Lee, D. Effect of anchors on the seismic performance of exposed column-base plate weak-axis connections. *J. Build. Eng.* **2020**, *32*, 101803. [CrossRef]
10. Yamanishi, T.; Tamai, H.; Takamatsu, T.; Matsuo, A. Elastic Rotational Rigidity of Exposed Column-base (In case of anchor bolt yield type). *J. Struct. Constr. Eng.* **2008**, *73*, 317–324. (In Japansese) [CrossRef]
11. Pan, J.; Huang, R.; Xu, J.; Wang, P.; Wang, Z.; Chen, J. Behavior of exposed column-base connections with four internal anchor bolts under seismic loading. *Structures* **2021**, *34*, 105–119. [CrossRef]
12. Tamai, H.; Ichinose, N. *On Deformation Capacity of Exposed Column-Base*; Graduate School of Engineering, Nagasaki University: Nagasaki, Japan, 2015; Volume 45. (In Japanese)
13. Lim, W.-Y.; Lee, D.; You, Y.-C. Exposed column-base plate strong-axis connections for small-size steel construction. *J. Constr. Steel Res.* **2017**, *137*, 286–296. [CrossRef]
14. Akiyama, H. *Seismic Design of Steel Column Bases*; GIHODO SHUPPAN Co., Ltd.: Tokyo, Japan, 1985. (In Japanese)
15. Yanagita, Y.; Hannuki, T.; Akiyama, H. *Experimental Study on Hysteretic Behaviors of Exposed Steel Column Bases: (Part III) Experimental Formulas of Week Base-Plate Types*; Summaries of Technical Papers of Annual Meeting Architectural Institute of Japan; SC-1, Structures III, Timber Structures Steel Structures Steel Reinforced Concrete Structures; Architectural Institute of Japan (AIJ): Tokyo, Japan, 2005; pp. 651–652. (In Japanese)
16. Cui, Y.; Wang, F.; Li, H.; Yamada, S. Rotational behavior of exposed column bases with different base plate thickness. *Steel Compos. Struct.* **2019**, *32*, 497–507. [CrossRef]
17. Kokubo, A.; Hannuki, T.; Yanagita, Y.; Akiyama, H. *Experimental Study on Hysteretic Behavior of Exposed Steel Column Bases: Part I Experimental Program and Results*; Summaries of Technical Papers of Annual Meeting Architectural Institute of Japan; C-1, Structures III, Timber Structures Steel Structures Steel Reinforced Concrete Structures; Architectural Institute of Japan (AIJ): Tokyo, Japan, 2004; pp. 721–722. (In Japanese)
18. Akiyama, H.; Yamada, S.; Takahashi, M.; Katsura, D.; Kimura, K.; Yahata, S. Full scale shaking table test of the exposed column bases. *J. Struct. Constr. Eng.* **1998**, *63*, 185–192. (In Japanese) [CrossRef] [PubMed]
19. Takamatsu, T.; Douki, H. Experimental Study on Restoring Force Characteristics of Exposed-Type Column Base. *Bull. Hiroshima Inst. Technol.* **2001**, *35*, 133–140. (In Japanese)
20. Inoue, K.; Suita, K. *Building Steel Structure's Theory and Design*; Kajima Institute Publishing Co., Ltd.: Tokyo, Japan, 2007. (In Japanese)

buildings

MDPI

Article

Effect of Diaphragm Above Concrete-Filled Part on Horizontal Load Capacity of Partially Concrete-Filled Circular Piers Subjected to Axial Forces

Yuki Chikahiro [1,*], Nguyen Thi Hong Huong [2] and Seiya Zenzai [3]

1 Department of Water Environment and Civil Engineering, Shinshu University, Wakasato 4-Chome, Nagano 380-8553, Japan
2 Shimizu Corporation, Kyobashi 2-16-1, Tokyo 104-8370, Japan
3 C.E. Management Integrated Laboratory Co., Ltd., Amenomiya 2347-3, Nagano 387-0001, Japan
* Correspondence: chikahiro@shinshu-u.ac.jp

Abstract: Partially concrete-filled steel tubes (PCFSTs) are often used to reduce the dead weight of concrete-filled steel tubes (CFSTs). Most previous studies have focused on the presence or absence of diaphragms directly above the concrete filling of PCFSTs, and few have focused on diaphragm characteristics. Therefore, this study presents the parametric analysis of partially concrete-filled steel tubes with circular cross-sections to clarify the effect of the diaphragm's parameters on the horizontal load capacity. The authors performed pushover analyses for a total of 84 cases, focusing on four axial force ratios, three diaphragm thicknesses, and seven diaphragm opening ratios. Although the thickness of the diaphragm had little effect on the horizontal load capacity, the opening ratio affected the horizontal load capacity. It was found that an opening ratio of 40–80% provided a higher horizontal load capacity than the 20%, 90%, and 95% openings.

Keywords: concrete-filled circular section piers; horizontal load capacity; diaphragm

1. Introduction

A concrete-filled steel tube (CFST) is a structure in which the inside of the steel tube is filled with concrete. Inside the CFST, the steel tube restrains the expansion of the concrete filling inside, resulting in a triaxial compressive stress state and providing a large load capacity. In civil engineering structures, it is widely used for steel piers of road bridges and elevated railway bridges to improve their seismic resistance because it enables rapid construction and construction in narrow areas [1]. In addition to column structures, there are also examples of CFSTs being used for girders and arch ribs [2–5]. In addition, partially concrete-filled steel tubes (PCFSTs) are often used to reduce the dead weight of concrete. For PCFSTs, it is recommended that a diaphragm be installed directly above the concrete filling to confine the concrete to prevent horizontal cracking of the concrete filling at the base of the steel tube and to allow the concrete filling to bear the axial force [6].

Many previous studies of PCFSTs with circular cross-sections have focused on design parameters, such as the concrete filling ratio, slenderness ratio, and diameter–thickness ratio [7,8] and finite element (FE) analyses have been performed [9–12]. However, few studies have focused on the diaphragm directly above the concrete filling in the PCFST. For example, Ota et al. studied the effect of concrete-filling repair for steel piers in relation to PCFSTs, focusing on the concrete filling height and the presence or absence of diaphragms [13].

The results showed that the installation of the diaphragm resulted in a higher maximum horizontal load, resulting in a larger displacement being required to reach the maximum horizontal load. In addition, Shimaguchi and Suzuki conducted experiments on the effect of concrete fill repair for steel piers with the PCFST, focusing on the filling ratio, radius thickness ratio, and the presence or absence of diaphragms [14]. This study reported that, when the piers have base strain values of up to 2% and a radius thickness ratio parameter of 0.065 to 0.080, restoration by simply filling with concrete is sufficient to restore load capacity. However, to mitigate damage when the base strain reaches approximately 5%, methods such as diaphragm installation are needed. Zenzai et al. also studied the effect of the diaphragm opening ratio on the seismic behavior of PCFSTs by FE analysis [15]. Seismic response analysis revealed that buckling occurred at the base of the PCFST or just above the diaphragm, depending on whether a diaphragm was present, and that these differences had a significant effect on the post-buckling behavior. As described above, most of the previous studies have focused on the presence or absence of diaphragms directly above the concrete filling of PCFSTs, and few have focused on diaphragm characteristics. The literature focusing on the diaphragm opening ratio [15] is limited to a total of eight cases, and further knowledge is needed to evaluate the effect of the diaphragm.

Therefore, this study can be regarded as a fundamental study of PCFSTs with circular cross-sections, and the influence of the diaphragm just above the concrete filling on the horizontal load capacity and buckling deformation was clarified with pushover analysis. Specifically, parametric analysis was performed by varying the axial compressive force acting directly above the steel tube in the PCFST in addition to the diaphragm thickness and opening ratio.

Section 2 explains the numerical conditions, including design parameters, FE models, material properties, boundary conditions, and contact conditions. Then, Section 3 shows the numerical results, horizontal load and horizontal displacement curve, shape deformation, and Mises stress distribution, focusing on the primary cases. Based on these results, Section 4 discusses the effects of the thickness and opening ratio of the diaphragm on the local buckling and horizontal load capacity of the PCFST, assessing the location of local buckling, the horizontal load capacity, and the Mises stress distribution of the PCFSTs. Finally, concluding remarks are presented in Section 5.

2. Numerical Conditions

2.1. Outline of Numerical Model

A summary of the numerical model is shown in Figure 1 and Table 1. The basic specifications of the steel tube were determined by referring to previous numerical studies by other authors [12], and the diameter D, thickness t_s, and height L were 900 mm, 9.3 mm, and 3850 mm, respectively. The dimensionless parameters related to the seismic performance of the steel tube, the radius thickness ratio parameter R_t and the slenderness ratio parameter λ_t, are expressed by the following equations [16] and set to 0.12 and 0.3, respectively:

$$R_t = \sqrt{3(1-{\mu_s}^2)}\frac{\sigma_y}{E_s}\frac{D}{2t} \quad (1)$$

$$\lambda_t = \frac{KL}{r}\frac{1}{\pi}\sqrt{\frac{\sigma_y}{E_s}} \quad (2)$$

where μ_s, σ_y, E_s, A_s, t, and r are the Poisson's ratio, yield stress, yielding stress of the steel, modulus of elasticity, cross-sectional area, thickness of the steel stube, and sectional secondary radius of the steel tube, respectively, and K is the effective buckling length coefficient of the steel tube ($K = 2.0$). These parameter settings are within the limits of specifications for highway bridges in Japan.

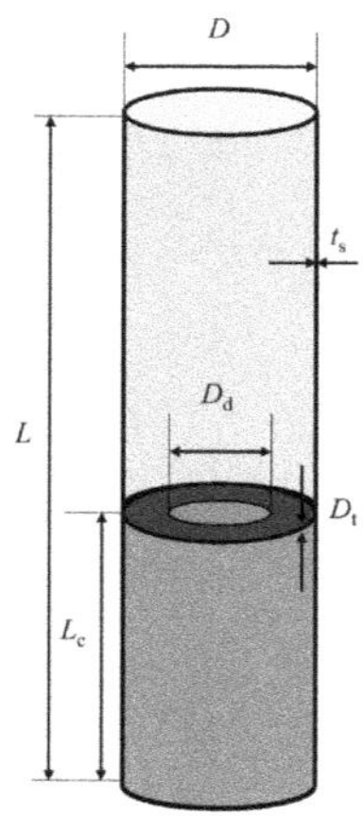

Figure 1. Numerical PCFST model.

Table 1. Numerical PCFST model specifications [12,17].

Parameter	Value
Diameter of steel tube D [mm]	900
Thickness of steel tube t_s [mm]	9.3
Height of steel tube L [mm]	3850
Section of steel tube A [mm^2]	26,023.4
Secondary radius of the section of steel tube r [mm]	314.9
Young's modulus of steel tube E_s [GPa]	206.0
Poisson's ratio of steel tube μ_s	0.3
Yielding stress of steel tube σ_y [MPa]	308.0
Concrete filling ratio L_c/L [%]	40
Radius thickness ratio parameter R_t	0.12
Slenderness ratio parameter λ_t	0.3
Opening ratio of diaphragm D_d/D [%]	20, 40, 50, 70 80, 90, 95, 100
Thickness of diaphragm D_t [mm]	4, 6, 12
Axial force ratio $n = N/N_y$ [%]	0, 10, 20, 30

In this study, a total of 84 cases were parametrically analyzed by varying the axial force ratio, diaphragm thickness, and opening ratio. Although the literature [16] stipulates that yielding should not be allowed in the unfilled portion of PCFSTs, the height of the concrete filling was set to 40% of the steel tube height L in all models, as it was positioned as a basic study of PCFSTs focusing on the diaphragm, as described in Section 1. The axial compressive force N acting on the top of the steel tube was set so that the axial force ratio n to the total cross-sectional yield axial force $N_y = \sigma_y A_s$ of the steel pipe cross-section, excluding the concrete filling, was 0%, 10%, 20%, and 30%, considering the actual design level. The diaphragms were welded directly on top of the concrete filling, and the opening ratio of the diaphragms was examined for 7 patterns ranging from 20% to 100% (without a diaphragm). The diaphragm opening ratio is the size of the diaphragm opening D_d divided by the diameter of the steel tube D. The thicknesses of the diaphragms were set to 6 mm and 12 mm, referring to previous studies [13,14,18] in which the diaphragm thickness was ranged from 67% to 120% of the thickness of the steel tube. A thinner thickness of 4 mm was also used for comparison and verification.

2.2. Elemental Partition

MSC's commercial FEM software (Marc2020) was used for the FE analysis. The mesh of the numerical model was discretized with three types of elements, as shown in Figure 2. Four-node shell elements were applied to the steel tube and diaphragms, and hexahedral solid elements were applied to the concrete filling. Above 90% of the column height, the number of elements was reduced by using two-node beam elements after confirming that the stresses do not exceed the elastic limit. The steel tube was divided into 90 sections in the circumferential direction of the steel tube. Elements were divided every 0.5% of the steel tube's height in the height direction for up to 60% of the column, and there were 15 divisions from 60% to 90% of the height. The diaphragm was divided into 90 circumferential sections, as in the steel tube. For modeling, an initial gap of 0.5 mm between the steel tube and diaphragm elements and the concrete elements was given, as carried out in reference [9], to account for the drying shrinkage of the concrete filling inside the tubes.

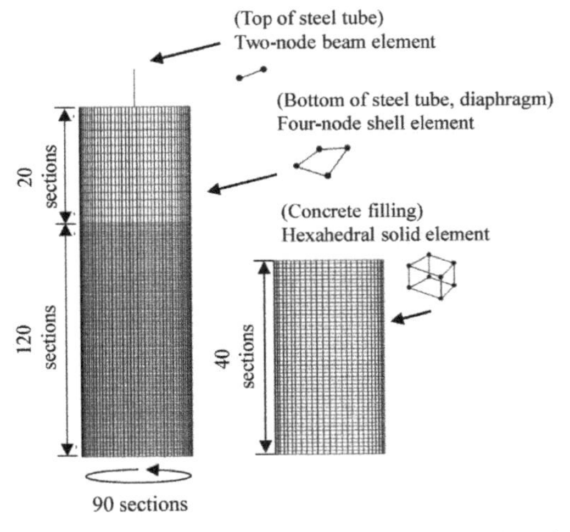

Figure 2. The sectioning of elements in the numerical model [12].

2.3. Material Properties

SS400 steel was used for the steel tubes and diaphragms, and the modulus of elasticity E_s, Poisson's ratio μ_s, and yield stress σ_y were assumed to be 206.0 GPa, 0.3, and 308.0 MPa, respectively. These properties were obtained from static cyclic loading tests on steel columns [17]. In addition, the stress–strain relationship shown in Figure 3 is of the curvilinear type with a yield shelf, as recommended in reference [6]. The yield condition of the steel tube was the von Mises yield condition, and the isotropic hardening law was applied. The concrete filling was assumed to be ordinary concrete with a uniaxial compressive strength of 22.0 MPa [19], and the modulus of elasticity and Poisson's ratio of the concrete were determined based on ACI Committee 318 [20]: $E_c = 4730$, $\sqrt{f_c} = 210$ GPa, and $\mu_c = 0.2$. In this study, the following Drucker–Prager rule was applied to define the modulus of elasticity of concrete:

$$f_c = \alpha I_1 + \sqrt{J_2} \tag{3}$$

where I_1 is the first invariant of the stress tensor, J_2 is the second invariant of the deviatoric stress tensor, and α is the material constant on which the hydrostatic stress depends ($\alpha = 0.2$ based on Balmer's experimental results [21]). The stress–strain relationship proposed by Popovics, shown in Figure 4, was used as the compressive stress–strain relationship [22]. Concrete cracking caused by tensile stress was treated as a contact problem between the concrete filling and the rigid surface of the base, assuming that the steel tube base, where the largest bending moment acts, is the location where concrete cracking occurs, as determined by Ngo et al. [23].

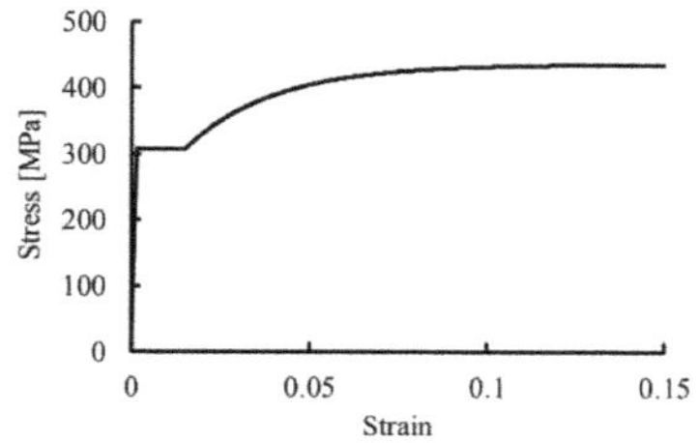

Figure 3. Steel tube stress–strain curve [12].

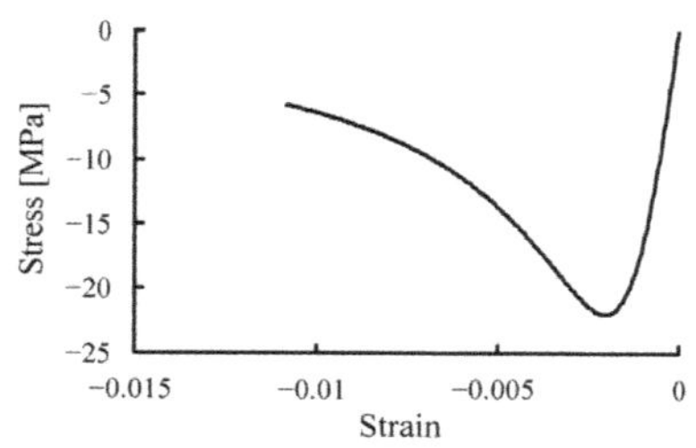

Figure 4. Compressive stress–compressive strain curve for concrete filling [12].

2.4. Boundary Conditions

The boundary conditions of the numerical model are shown in Figure 5. The base of the column was assumed to be completely fixed, and the head of the column was subjected to an axial compressive force N and horizontal displacement δ, representing the mass of the superstructure. In the FE analysis, the pushover analysis was conducted by loading the axial compressive force N firstly, and then applying the horizontal displacement δ. The horizontal yield displacement δ_y applied to the column head is calculated by the yield load H_y of the steel tube and is expressed as follows:

$$\delta_y = \frac{H_y}{3E_sI}L^3 \tag{4}$$

$$H_y = \left(\sigma_y - \frac{N}{A_s}\right)\frac{I}{Ly} \tag{5}$$

where y is the distance from the center of the steel tube to the outside. In this study, as the axial force ratio changed to 0%, 10%, 20%, and 30%, the yield displacement δ_y was 16.4 mm, 14.8 mm, 13.2 mm, and 11.5 mm, respectively.

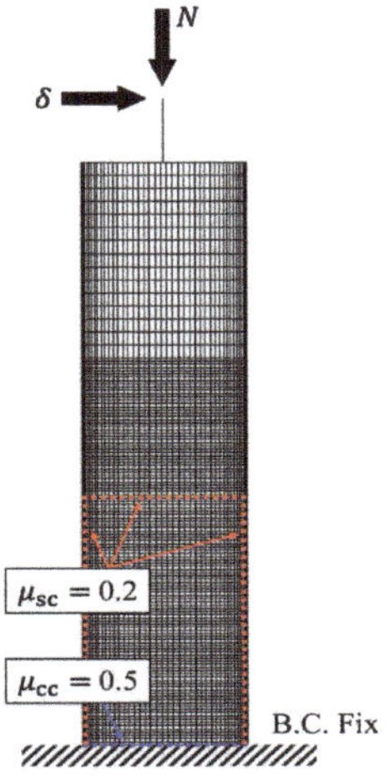

Figure 5. Numerical model boundary conditions.

2.5. Contact Conditions

When concrete-filled steel tube columns are subjected to horizontal forces, contact, delamination, and friction occur between the steel tube and the concrete filling. In this study, the extended Lagrange multiplier method was used for these contact problems. The Coulomb friction model shown in Figure 6 was used for frictional behavior. The vertical axis in the figure indicates the shear stress τ and the horizontal axis u_t indicates the relative displacement between the steel tube and concrete. In the Coulomb friction model, the shear stress is bonded to the interface until the maximum static frictional stress τ_{cr} is reached, at which point the interface begins to slip. The maximum static friction force τ_{cr} at is given by

$$\tau_{cr} = \mu p \tag{6}$$

where μ is the coefficient of friction and p is the normal stress. A coefficient of friction of μ_{sc} = 0.2 was used for the friction between the steel tube and concrete and between the diaphragm and concrete, based on the work of Johansson et al. [24], and a coefficient of friction of μ_{cc} = 0.5 was used between the concrete filling and the base. The numerical method used in this study is based on the literature [15]. Here, the accuracy verification was performed up to a horizontal displacement of approximately 120 mm for the historical curves obtained in experiment [17] under cyclic loading.

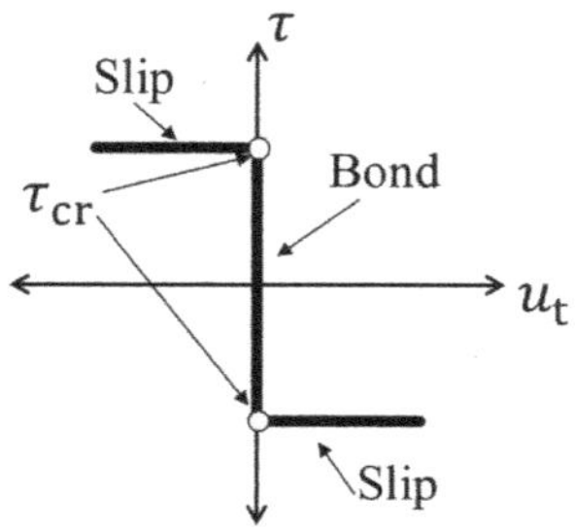

Figure 6. Coulomb's friction law [12].

3. Numerical Results

To show the typical behavior of a PCFST, we first focus on the numerical results for a diaphragm thickness of 6 mm and a 50% opening ratio.

3.1. Relationship Between Horizontal Load and Horizontal Displacement

Figure 7 shows load–displacement curves for each axial force ratio n. The vertical and horizontal axes represent the horizontal load and horizontal displacement of the steel tube head, respectively. The model follows the same path regardless of the axial force ratio up to around 600 kN, but the model with an axial force ratio of 30% loses strength after showing a horizontal load capacity of approximately 744 kN. Local buckling occurs just above the concrete filling. As shown in Section 3.2, the models differ in terms of the location of out-of-plane deformation, but the loss of bearing capacity in the load–displacement curves indicates that out-of-plane deformation developed just above the concrete filling, leading to local buckling.

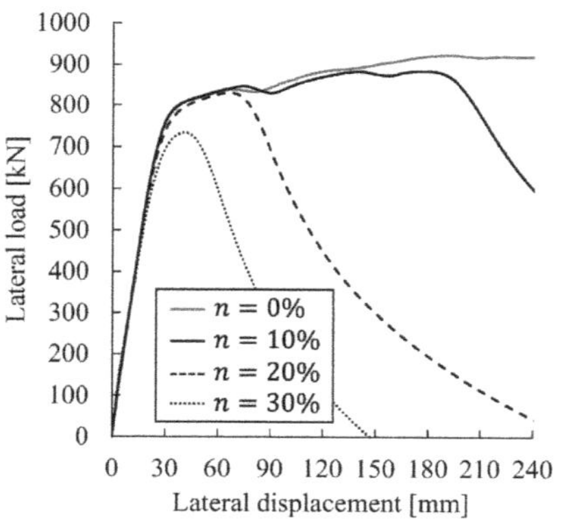

Figure 7. Horizontal load–horizontal displacement curves of PCFSTs with diaphragm thickness of 6 mm and 50% opening ratio.

Figure 8 shows the horizontal displacements from $1\delta_y$ to $10\delta_y$, focusing on the length from the base of the steel tube, which is on the compression side in the loading direction, up to a height of 2000 mm. The gray line in the figure indicates the location of the diaphragm in the steel tube. In Figure 8a, where the axial force ratio is 0%, the out-of-plane deformation at the base of the steel tube is larger than that directly above the concrete filling. As the axial force ratio increases to 10%, 20%, and 30%, the out-of-plane deformation at the base of the steel tube becomes smaller and that directly above the concrete filling becomes larger.

In Figure 8c with an axial force ratio of 20%, the diaphragm oscillates from around $6\delta_y$, and in Figure 8d with an axial force ratio of 30%, the diaphragm oscillates from around $4\delta_y$. The horizontal displacement between the installation position of the diaphragm and the base of the steel tube, approximately 1700 mm from the base of the tube, is almost constant. Considering the results in Figure 7, this trend is likely due to local buckling.

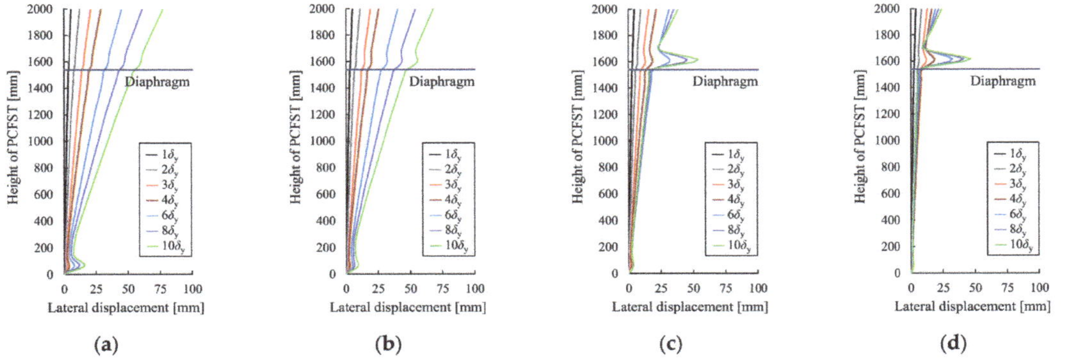

Figure 8. Horizontal displacement on compression side of steel tube of PCFST with diaphragm thickness of 6 mm and 50% opening ratio: (**a**) $n = 0\%$; (**b**) $n = 10\%$; (**c**) $n = 20\%$; (**d**) $n = 30\%$.

3.2. Shape Deformation and Mises Stress Distribution State

Mises stress diagrams for steel tubes at 0%, 10%, 20%, and 30% axial force ratios at $6\delta_y$ yield displacement are shown in Figure 9. The red arrows in the figure indicate the locations where out-of-plane deformation developed as a result of loading. The figure shows that out-of-plane deformation occurs at the base of the steel tube when the axial force ratio is 0%, directly above the base of the steel tube and concrete filling when the axial force ratio is 10% or 20%, and directly above the concrete filling when the axial force ratio is 30%, and that these developments affect the trend of the load–displacement curve.

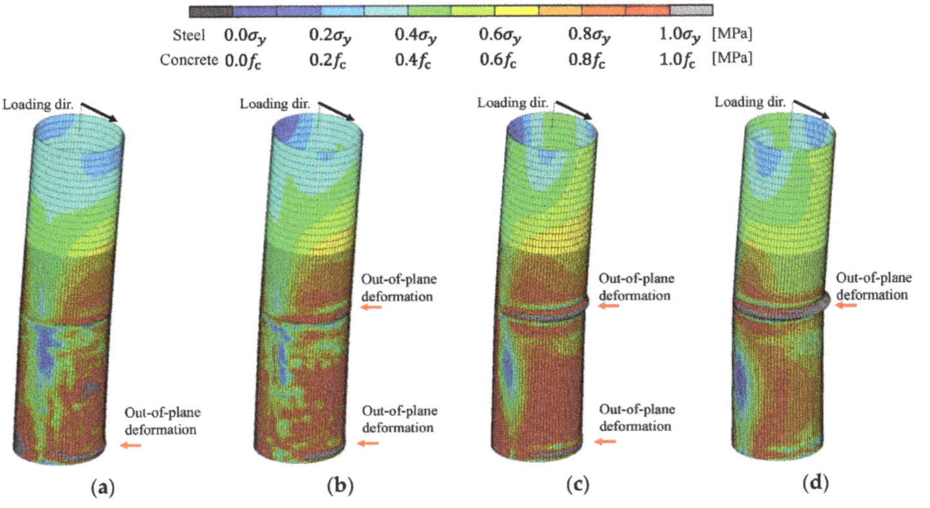

Figure 9. Mises stress diagram of PCFST with horizontal displacement of $6\delta_y$ and 50% opening ratio (deformation factor of 3 times): (**a**) $n = 0\%$; (**b**) $n = 10\%$; (**c**) $n = 20\%$; (**d**) $n = 30\%$.

Figures 10–13 focus on Mises stresses at the diaphragm inside the PCFST and at the top of the concrete filling from the conditions shown in Figure 9. The black arrows in the figures indicate the loading direction, and the black solid and dashed lines indicate the position corresponding to the width of the diaphragm at 50% and 80% opening ratios, respectively. When the axial force ratio was 20%, high stresses also occurred in the center of the diaphragm, but in general the stresses were greater at the periphery than at the 80% opening ratio indicated by the dashed line. The same was true for the concrete filling, which tended to exhibit less stress in the center and greater stress closer to the periphery.

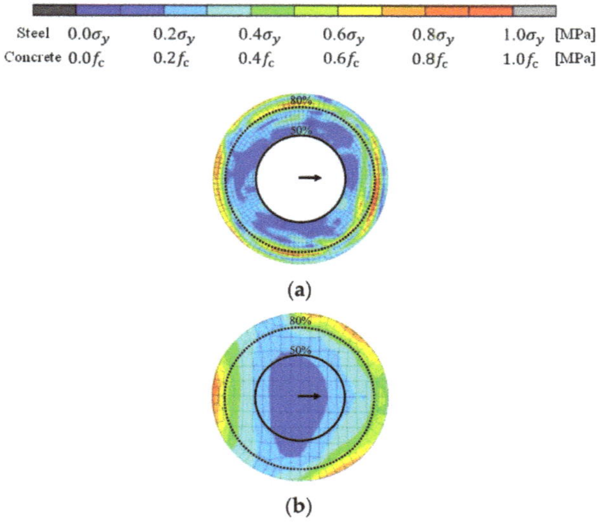

Figure 10. Mises stress diagram at n = 0%: (**a**) diaphragm; (**b**) top of concrete filling.

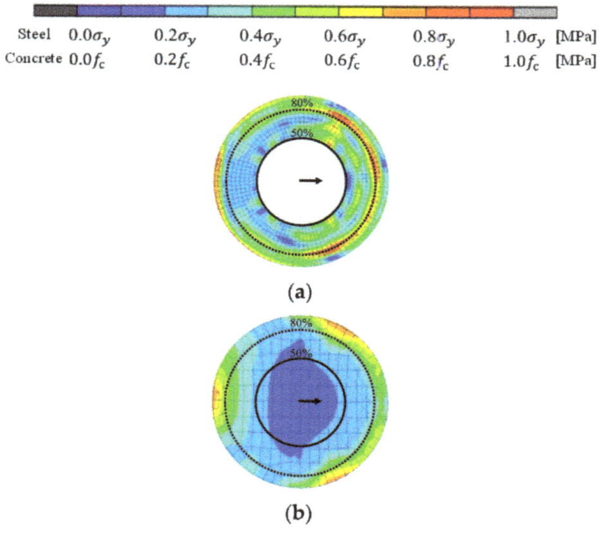

Figure 11. Mises stress diagram at n = 10%: (**a**) diaphragm; (**b**) top of concrete filling.

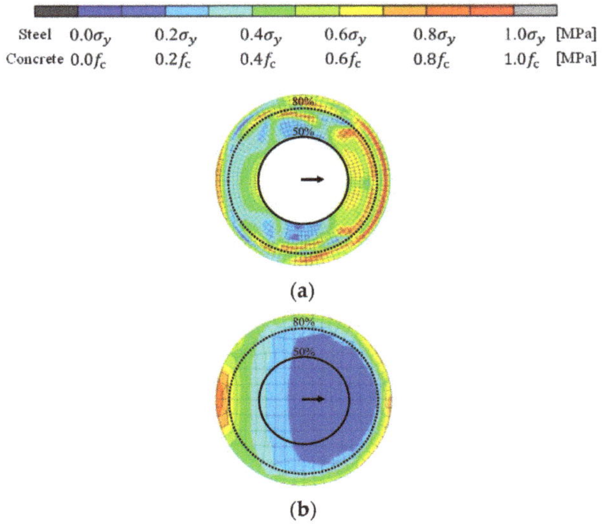

Figure 12. Mises stress diagram at $n = 20\%$: (**a**) diaphragm; (**b**) top of concrete filling.

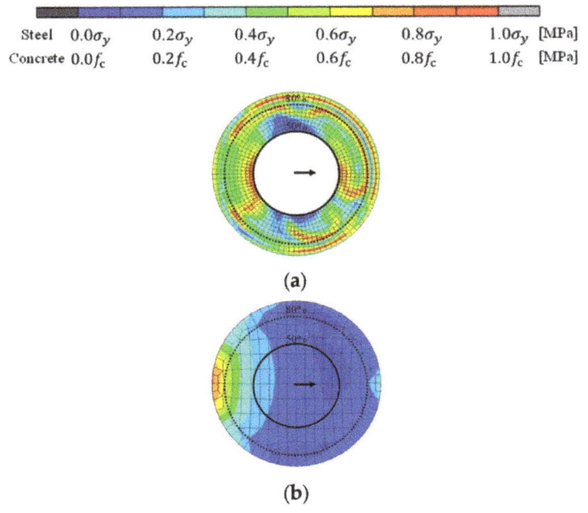

Figure 13. Mises stress diagram at $n = 30\%$: (**a**) diaphragm; (**b**) top of concrete filling.

4. Discussion Focusing on the Diaphragm

Here, results presented in Section 3 are used to discuss the effects of the thickness and opening ratio of the diaphragm on the local buckling and horizontal load capacity of the PCFST.

4.1. Location of Local Buckling

4.1.1. Primary Cases

Figure 14 shows the horizontal displacement on the compression side of a steel tube with a diaphragm thickness of 6 mm when subjected to an axial force ratio of 10%. The coordinate axes and legend in the figure are the same as in Figure 8.

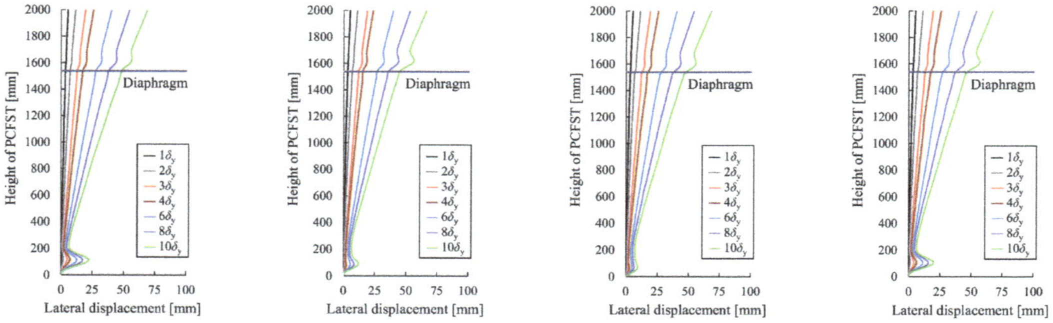

(a) (b) (c) (d)

Figure 14. Horizontal displacement on compression side of steel tube of PCFST at axial force ratio of 10% and diaphragm thickness of 6 mm: (**a**) 20% opening ratio: (**b**) 50% opening ratio; (**c**) 80% opening ratio; (**d**) 95% opening ratio.

In Figure 14a,d, the out-of-plane deformation at the base of the steel tube is greater than that directly above the concrete. In contrast, in Figure 14b, the out-of-plane deformation at the base of the steel tube and directly above the concrete is almost the same, while in Figure 14c, the out-of-plane deformation directly above the concrete is larger than that at the base of the steel tube, with different trends emerging depending on the opening ratio of the diaphragm. From this, it can be inferred that the location of local buckling may vary depending on the diaphragm opening ratio.

4.1.2. All Cases

Table 2 summarizes the buckling positions of the PCFSTs obtained from the parametric analysis. The buckling that occurred at the base of the steel tube is colored black, the buckling that occurred at two locations directly above the concrete filling is colored white, and the buckling that occurred at two locations, one at the base of the steel tube and the other directly above the concrete filling, is colored gray. Here, the timing at which the displacement of the out-of-plane deformation of the steel pipe increased rapidly was defined as the position of local buckling.

Table 2. Classification of locations where local buckling occurred.

D_t		4 mm				6 mm				12 mm			
n		0%	10%	20%	30%	0%	10%	20%	30%	0%	10%	20%	30%
D_d/D	20%	[black]	[black]	[gray]	[gray]	[black]	[black]	[gray]	[gray]	[black]	[black]	[gray]	[gray]
	40%	[black]	[gray]	[gray]	[white]	[black]	[gray]	[white]	[white]	[black]	[gray]	[white]	[white]
	50%	[black]	[gray]	[white]	[white]	[black]	[gray]	[white]	[white]	[black]	[gray]	[white]	[white]
	70%	[black]	[gray]	[white]	[white]	[black]	[gray]	[white]	[white]	[black]	[white]	[white]	[white]
	80%	[black]	[white]	[white]	[white]	[black]	[white]	[white]	[white]	[black]	[white]	[white]	[white]
	90%	[gray]	[white]	[white]	[white]	[gray]	[white]	[white]	[white]	[gray]	[white]	[white]	[white]
	95%	[black]	[black]	[gray]	[white]	[black]	[gray]	[gray]	[white]	[black]	[gray]	[white]	[white]
	100%	[black]	[black]	[black]	[black]	[black]	[black]	[black]	[black]	[black]	[black]	[black]	[black]

The table shows that local buckling occurred at the base of the steel tube in most models when the axial force ratio was 0%, but as the axial force increased, the location of local buckling changed from two locations, one at the base of the steel tube and the other directly above the concrete filling, to only directly above the concrete filling. At an axial force ratio of 30%, most models experienced local buckling just above the concrete filling. Based on the results presented in Section 3.1, it should be noted that if local buckling only occurs directly above the concrete filling, the load capacity may be rapidly lost. The results for different diaphragm opening ratios showed that more local buckling occurred at the base of the steel tube at 20% and 95% opening ratios than at other ratios, while at 80% and 90% opening ratios, local buckling only occurred directly above the concrete filling. In contrast, for the plate thickness, local buckling occurred at the base of the steel tube with a 95% opening ratio and a plate thickness of 4 mm. The local buckling occurred directly above the concrete filling in some cases when the plate was as thick as 12 mm.

4.2. Horizontal Load Capacity

4.2.1. Primary Cases

Figure 15 shows load–displacement curves for each thickness of diaphragm. The vertical and horizontal axes represent the horizontal load and horizontal displacement of the steel tube head, respectively.

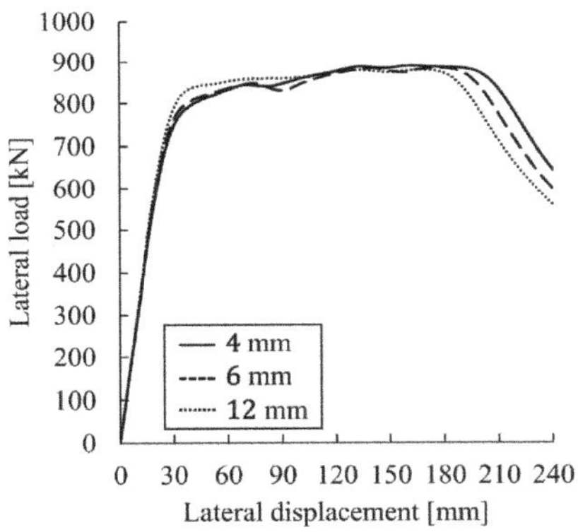

Figure 15. Horizontal load–horizontal displacement curves of PCFSTs with axial force ratio 10% and 50% opening ratio.

The model follows the same path regardless of the thickness of the diaphragm up to around 600 kN. There was no significant difference in the curve trend and horizontal load capacity for each thickness of diaphragm, even after the yield displacement, and only the timing of load drop due to local buckling that occurred directly above the concrete filling varied with the thickness of diaphragm. Such local buckling occurred more quickly in the model with the thicker diaphragm. These results suggest that there was little difference in the confinement effect of the concrete filling by the diaphragm, but the thicker diaphragm caused the concrete filling to act like the base, and thus local buckling was more likely to occur directly above the concrete filling.

4.2.2. All Cases

Figure 16 shows the relationship between the horizontal load capacity and opening ratio for different axial force ratios and diaphragm thicknesses. Figure 16a shows the overall view, and Figure 16b–d show detailed views of the axial force ratios of 0%, 10%, and 20%. The empty circle (○) indicates buckling that only occurred directly above the concrete filling, while the filled circle (•) indicates buckling at the steel tube base and at two locations (the steel tube base and directly above the concrete filling). Furthermore, the red circles (•) indicate the highest horizontal load capacity at each thickness of the diaphragm for each axial force ratio, and the red arrows (↓) indicate the maximum value, circled in red. The blue circles (•) in the figure indicate the opening ratio of the diaphragm with the lowest horizontal load capacity at each diaphragm thickness for each axial force ratio (excluding the 100% opening ratio), and the blue arrows (↑) indicate the minimum value, circled in blue.

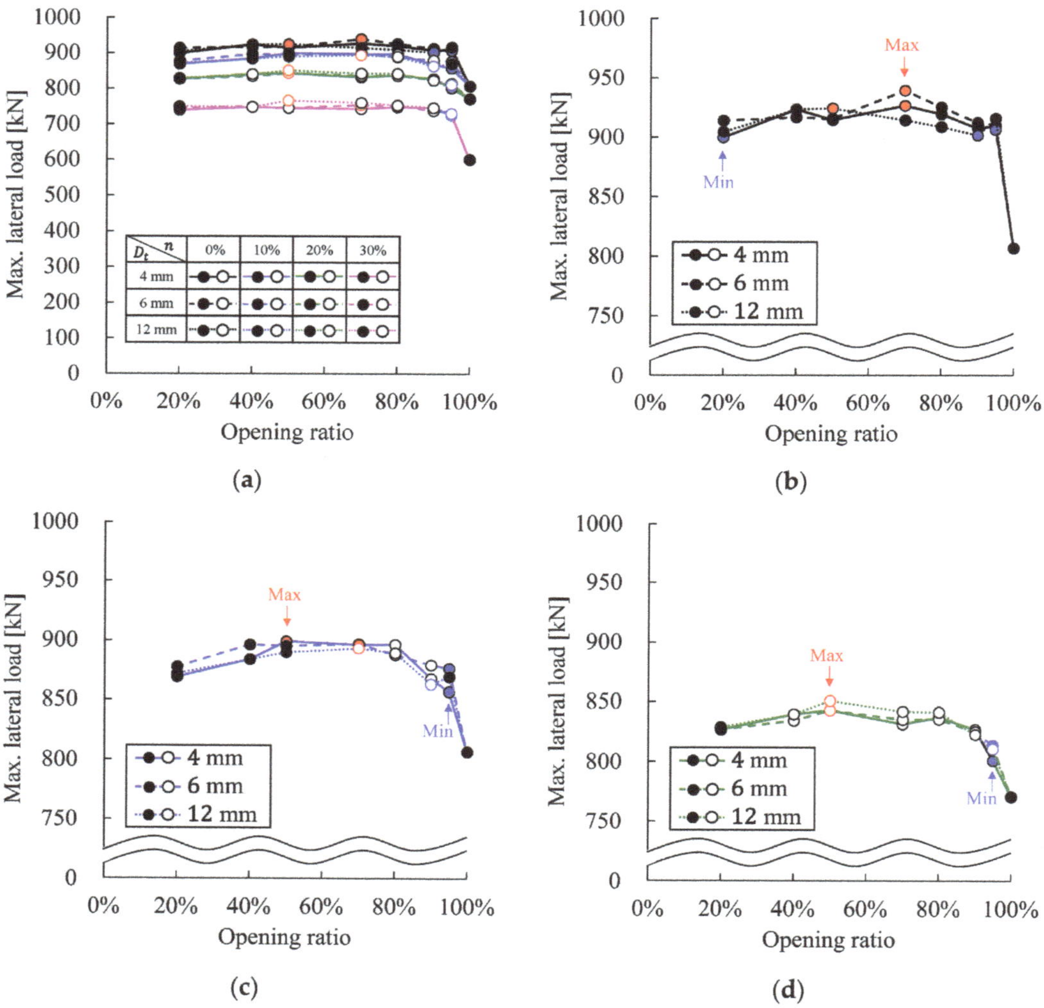

Figure 16. Horizontal load capacity–opening ratio relationship: (**a**) overview; (**b**) detail of axial ratio of 0%; (**c**) detail of axial ratio of 10%; (**d**) detail of axial ratio of 20%.

Figure 16a shows that the horizontal load capacity of the PCFST decreases as the axial force ratio increases. In the case of CFST structures, it is known that the effective cross-sectional area of the concrete filling increases up to 20% of the axial force ratio, resulting in an increase in bending capacity and an increase in horizontal capacity. However, the PCFST structure may be less affected than the CFST structure because the out-of-plane deformation propagates more easily just above the diaphragm due to the lower concrete filling rate. Opening ratios that exhibited the maximum horizontal load capacity varied between 40% and 70% for each plate thicknesses, from 50% to 70% for a 0% axial force ratio, from 50% to 70% for a 10% axial force ratio, from 50% for a 20% axial force ratio, and from 40% to 70% for a 30% axial force ratio. The minimum horizontal load capacity occurred mostly for 90% to 95% opening ratios, except for the 100% opening ratio, where no diaphragm was installed. This is a trend similar to that described in Section 3.2, and it indicates that an extremely large opening ratio does not provide enough contact area to transmit the load sufficiently to the concrete filling. Figure 16b–d show that the maximum horizontal load capacity for each thickness of the diaphragm at each axial force ratio and the minimum value, excluding the opening ratio of 100% (no diaphragm), have errors of 2.5% to 3.5% with respect to the maximum value for an axial force ratio of 0%, 2.3% to 4.7% for an axial force ratio of 10%, and 3.4% to 5.0% for an axial force ratio of 20%. Therefore, the effect of the diaphragm opening ratio on the maximum horizontal load capacity was not as large as that of the plate thickness and was found to be limited to an error of a few percent.

4.3. Mises Stress Distribution Diagram

To explore the effect of the opening ratio of the diaphragm, stress diagrams for PCFSTs with an axial force ratio of 10% and diaphragm thickness of 6 mm at horizontal displacements of $1\delta_y$, $6\delta_y$, and $10\delta_y$ are shown in Figure 17. Here, δ_y is the yield displacement of the steel tube. The PCFSTs are viewed from the compression side with respect to the loading direction. Top views of the concrete filling are also shown.

For a horizontal displacement of $1\delta_y$, the range of high Mises stresses at the base of the steel tube decreased and the range of high Mises stresses at the top of the concrete filling increased as the opening ratio increased from 20% to 50% to 80%. In contrast, when the opening ratio was 95%, the distribution of Mises stresses for both the steel tube and concrete filling was closer to that of the 20% opening ratio than to that of the 50% and 80% opening ratios. However, as the horizontal displacement increased to $6\delta_y$, the Mises stress diagram of the concrete filling showed a similar distribution regardless of the opening ratio. This suggests that the effect of the initial difference is small. In contrast, at a horizontal displacement of $10\delta_y$, the Mises stress just below the diaphragm with a 95% opening ratio was lower than that of the 20%, 50%, and 80% ratios, indicating a lower burden on the concrete filling. The Mises stress diagram of the diaphragm shown in Figure 18 indicates that the entire area of the diaphragm reached yield stress only when the opening ratio was 95%. As shown in Table 2, the tendency for local buckling to occur at the base of the steel tube when the opening ratio is 95% may be due to the diaphragm yielding. Local buckling at the base of the steel tube decreased as the thickness increased, even when the opening ratio was the same at 95%, probably because the thicker plate delayed the yielding of the diaphragm.

These results suggest that the risk of local buckling directly above the concrete filling can be reduced when the diaphragm thickness is thin and the opening ratio is small within the scope of this study. Even when the opening ratio is extremely small (95%), local buckling is likely to occur at the base of the steel tube due to the area of the diaphragm and its yielding. However, because the diaphragms in this study were examined under a limited filling ratio, radius thickness ratio, and slenderness ratio, the relevance of these parameters should be examined in future studies.

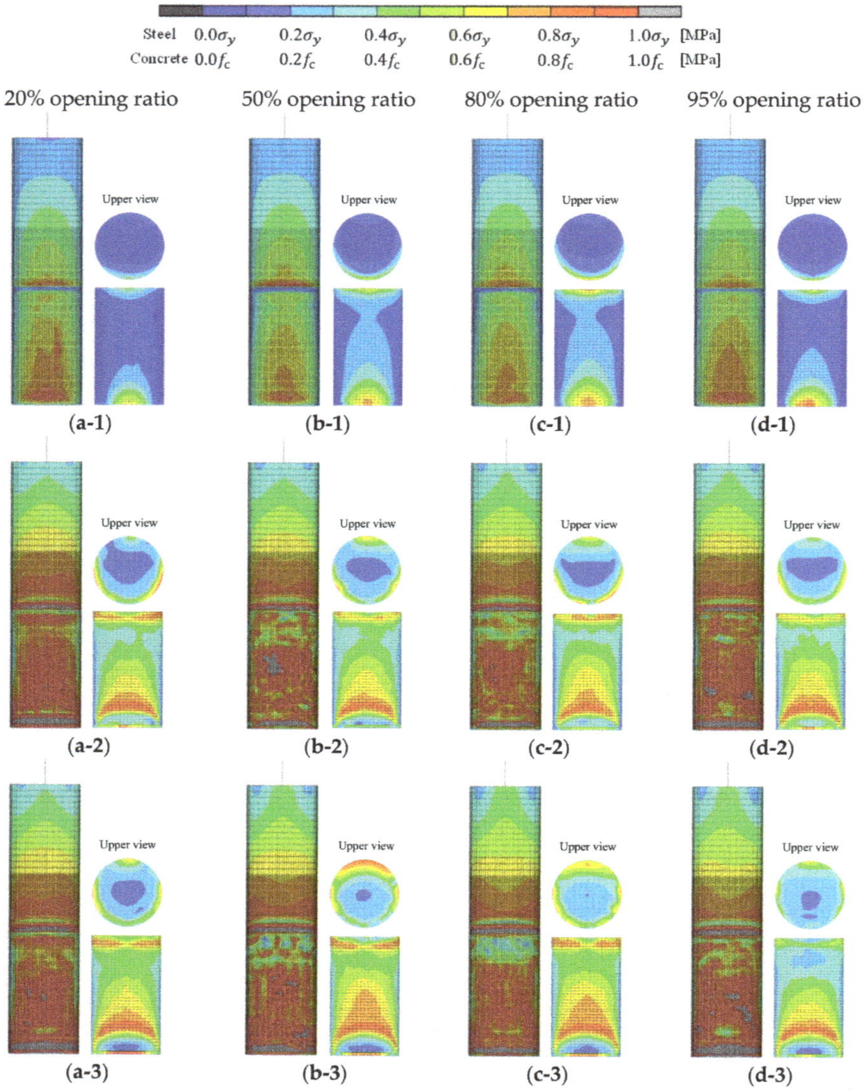

Figure 17. Mises stress diagram of steel tube and filled in concrete with axial force ratio 10% and diaphragm thickness 6 mm: (**a-1**) $1\delta_y$ with 20% opening ratio; (**a-2**) $6\delta_y$ with 20% opening ratio; (**a-3**) $10\delta_y$ with 20% opening ratio; (**b-1**) $1\delta_y$ with 50% opening ratio; (**b-2**) $6\delta_y$ with 50% opening ratio; (**b-3**) $10\delta_y$ with 50% opening ratio; (**c-1**) $1\delta_y$ with 80% opening ratio; (**c-2**) $6\delta_y$ with 80% opening ratio; (**c-3**) $10\delta_y$ with 80% opening ratio; (**d-1**) $1\delta_y$ with 95% opening ratio; (**d-2**) $6\delta_y$ with 95% opening ratio; (**d-3**) $10\delta_y$ with 95% opening ratio.

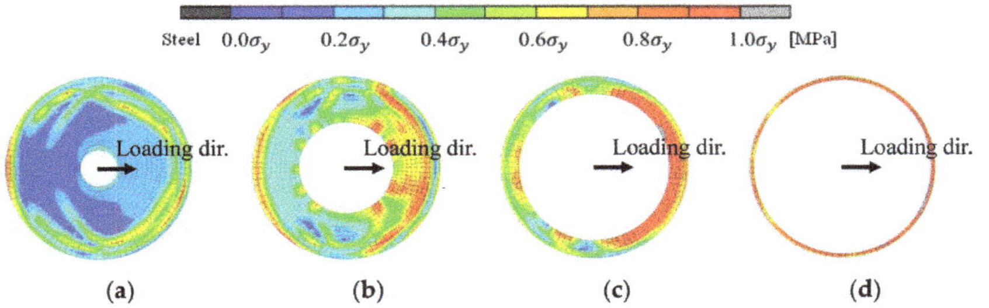

Figure 18. Mises stress diagram of diaphragm at horizontal displacement of $10\delta_y$ with axial force ratio of 10% and diaphragm thickness of 6 mm: (**a**) 20% opening ratio; (**b**) 50% opening ratio; (**c**) 80% opening ratio; (**d**) 95% opening ratio.

5. Conclusions

Most of the previous PCFSTs's studies have focused on the presence or absence of diaphragms, and few have focused on the diaphragms' characteristics. Therefore, this study can be regarded as a fundamental study of PCFSTs with circular cross-sections, and the influence of the diaphragm just above the concrete filling on the horizontal load capacity and buckling deformation was clarified with pushover analysis. The authors performed pushover analyses for a total of 84 cases, focusing on four axial force ratios, three diaphragm thicknesses, and seven diaphragm opening ratios.

The findings of this study are summarized below.

- The thickness of the diaphragm had little effect on the horizontal load capacity, and the amount of variation in the horizontal load capacity with respect to changes in the axial force ratio was also small.
- The maximum horizontal load capacity was distributed from 40% to 70% opening ratios, and minimum values occurred from 90% to 95%, except for the 100% opening ratio, where no diaphragm was installed.
- When the opening ratio of the diaphragm was as small as 20%, local buckling was likely to occur at the base of the steel tube, and when the opening ratio was as large as 80% or 90%, local buckling was only likely to occur directly above the concrete filling.
- When the opening ratio of the diaphragm was as large as 95%, local buckling was likely to occur at the base of the steel tube due to the smaller contact area between the diaphragm and the concrete filling and the yielding of the diaphragm.

Author Contributions: Conceptualization, Y.C.; methodology, Y.C.; software, N.T.H.H.; validation, S.Z.; formal analysis, N.T.H.H.; data curation, S.Z. and N.T.H.H.; writing—original draft preparation, Y.C. and N.T.H.H.; writing—review and editing, Y.C. and S.Z.; project administration, Y.C. All authors have read and agreed to the published version of the manuscript.

Funding: This research received no external funding.

Data Availability Statement: The datasets presented in this article are not readily available because the data are part of an ongoing study or due to technical limitations.

Conflicts of Interest: Author Nguyen Thi Hong Huong was employed by Shimizu Corporation. Author Seiya Zenzai was employed by C.E. Management Integrated Laboratory Co., Ltd. The remaining author declares that the research was conducted in the absence of any commercial or financial relationships that could be construed as a potential conflict of interest.

References

1. Han, L.H.; Li, W.; Bjorhovde, R. Developments and advanced applications of concrete-filled steel tubular (CFST) structures: Members. *J. Constr. Steel Res.* **2014**, *100*, 211–228. [CrossRef]
2. Zhang, D.Y.; Li, X.; Yan, W.M.; Xie, W.C.; Pandey, M.D. Stochastic seismic analysis of a concrete-filled steel tubular (CFST) arch bridge under tridirectional multiple excitations. *Eng. Struct.* **2013**, *52*, 355–371. [CrossRef]
3. Cho, J.; Moon, J.; Ko, H.J.; Lee, H.E. Flexural strength evaluation of concrete-filled steel tube (CFST) composite girder. *J. Constr. Steel Res.* **2018**, *151*, 12–24. [CrossRef]
4. Chen, B.W.; Han, L.H.; Qin, D.Y.; Li, W. Life-cycle based structural performance of long-span CFST hybrid arch bridge: A study on arch of Pingnan Third Bridge. *J. Constr. Steel Res.* **2023**, *207*, 107939. [CrossRef]
5. Hu, W.; Liu, J.; Liu, Y.; Lyu, Y.; Li, H.; Xian, J.; Ma, Y. Construction alignment and closure control of CFST truss arch bridges based on temperature effect. *Structures* **2024**, *63*, 106471. [CrossRef]
6. JSCE/Committee on Steel Structure. *Guidelines for Stability Design of Steel Structures*, 2nd ed.; Japan Society of Civil Engineers: Tokyo, Japan, 2005; pp. 428–429. (In Japanese)
7. Iura, M.; Orino, A.; Ishizawa, T. Elasto-Plastic Behaviour of Concrete-Filled Steel Tubular Columns. *J. JSCE* **2002**, *696*, 285–298. (In Japanese) [CrossRef] [PubMed]
8. Orino, A.; Iura, M. Tests on concrete-filled steel tubular columns subject to lateral load. In Proceedings of the 3rd Symposium on Nonlinear Numerical Analysis and Application to Seismic Design of Steel Structures, Nagoya, Japan, 7–8 January 2000; pp. 83–88. (In Japanese).
9. Goto, Y.; Kumar, G.P.; Kawanishi, N. FEM analysis for hysteretic behavior of CFT bridge piers considering interaction between steel tube and infilled concrete. *J. JSCE A* **2009**, *65*, 487–504. (In Japanese) [CrossRef]
10. Goto, Y.; Kumar, G.P.; Kawanishi, N. Nonlinear finite-element analysis for hysteretic behavior of thin-walled circular steel columns with in-filled concrete. *J. Struct. Eng.* **2010**, *136*, 1413–1422. [CrossRef]
11. Goto, Y.; Mizuno, K.; Kumar, G.P. Nonlinear finite element analysis for cyclic behavior of thin-walled stiffened rectangular steel columns with in-filled concrete. *J. Struct. Eng.* **2012**, *138*, 571–584. [CrossRef]
12. Zenzai, S.; Chikahiro, Y.; Shimizu, S. Estimation Equation for Horizontal Load Bearing Capacity of Circular PCFST with Diaphragm. *Appl. Sci.* **2022**, *12*, 8739. [CrossRef]
13. Ota, T.; Suzuki, M.; Shimaguchi, Y. A Study on Seismic Performance of Circular Steel Bridge Piers which have Different Level of Damages and Concrete Filled Repair. *J. JSCE A2 (Appl. Mech.)* **2013**, *69*, 381–390. (In Japanese) [CrossRef] [PubMed]
14. Shimaguchi, Y.; Suzuki, M. Seismic performance evaluation of circular steel bridge piers which have damage and concrete filled repair. *J. Struct. Eng.* **2015**, *61A*, 292–301. (In Japanese) [CrossRef]
15. Zenzai, S.; Shimizu, S.; Chikahiro, Y.; Ohkami, T. Effect of Diaphragm on Seismic Behaviour of Partially Concrete-Filled Steel Tube. *Steel Constr. Eng.* **2019**, *26*, 25–36. (In Japanese) [CrossRef]
16. Japan Road Association. *Specification for Highway Bridges Part V: Seismic Design*; Japan Road Association: Tokyo, Japan, 2012; pp. 251–274.
17. Public Works Research Institute, Ministry of Construction. *Cooperative Research on a Limit State Design of Piers of Highway Bridge Under Seismic Loading (Summarized Edition)*; Public Works Research Institute, Ministry of Construction: Tsukuba, Japan, 1999; p. 51. (In Japanese)
18. Goto, Y.; Seki, K.; Ebisawa, T.; Lu, X. Improved seismic performance of CFT piers by a newly discovered mechanism to restrain progress of local buckling deformations. *J. JSCE A1 (Struct. Eng. Earthq. Eng.)* **2013**, *69*, 101–120. (In Japanese) [CrossRef]
19. Yuan, H.; Dang, J.; Aoki, T. Behavior of partially concrete-filled steel tube under bidirectional seismic excitations. *J. Constr. Steel Res.* **2014**, *93*, 44–54. [CrossRef]
20. *ACI318-99*; Building Code Requirement for Structural Concrete and Commentary. American Concrete Institute: Indianapolis, IN, USA, 1999; p. 83.
21. Chen, W.F. *Plasticity in Reinforced Concrete*; J. Ross Publishing LLC: New York, NY, USA, 2007; pp. 19–48.
22. Popovics, S. A Numerical Approach to the Complete Stress-Strain Curve of Concrete. *Cem. Concr. Res.* **1973**, *3*, 583–599. [CrossRef]
23. Ngo, D.; Scordelis, A.C. Finite Element Analysis of Reinforced Concrete Beams. *J. Proc.* **1967**, *64*, 152–163. [CrossRef]
24. Johansson, M.; Gylltoft, K. Structural behavior of slender circular steel-concrete composite columns under various means of load application. *Steel Compos. Struct.* **2001**, *1*, 393–410. [CrossRef]

buildings

MDPI

Article

Seismic Optimization of Fluid Viscous Dampers in Cable-Stayed Bridges: A Case Study Using Surrogate Models and NSGA-II

Qunfeng Liu [1,*], Zhen Liu [1], Jun Zhao [2], Yuhang Lei [1], Shimin Zhu [3] and Xing Wu [3]

1 School of Architecture and Civil Engineering, Xi'an University of Science and Technology, Xi'an 710054, China; 22204053031@stu.xust.edu.cn (Z.L.); leiyuhanglyh@163.com (Y.L.)
2 School of Science, Xi'an University of Architecture and Technology, Xi'an 710055, China; zjbridge@xauat.edu.cn
3 CCCC First Highway Consultants Co., Ltd., Xi'an 710068, China; shimin_zhu@outlook.com (S.Z.); wuxing.cccc@outlook.com (X.W.)
* Correspondence: qliu@xust.edu.cn

Abstract: This study investigates two optimization strategies to enhance the seismic performance of cable-stayed bridges equipped with Fluid Viscous Dampers (FVDs). A detailed finite element model of a case study bridge was developed to evaluate the effectiveness of these strategies in optimizing FVD parameters for seismic mitigation. The first strategy employs a traditional parametric analysis approach, which identifies optimal parameters by examining their influence on seismic performance. The second strategy employs a data-driven surrogate model, specifically an Artificial Neural Network (ANN), integrated with the NSGA-II optimization algorithm. This surrogate model significantly reduced computational demands during the optimization process, offering a more efficient and scalable solution for the optimization process. Results demonstrate that the ANN-based approach effectively addresses multi-objective optimization challenges while providing a robust framework for improved seismic performance in cable-stayed bridges. These findings highlight the potential of the ANN-based strategy in the seismic optimization of FVD parameters for cable-stayed bridges.

Keywords: cable-stayed bridge; fluid viscous damper; parameter optimization; surrogate model; NSGA-II

1. Introduction

Cable-stayed bridges are widely utilized in modern long-span and landmark bridge engineering due to their excellent structural performance, notable spanning capacity, and aesthetic appeal. Based on the connection configurations between pylons and decks, cable-stayed bridges are categorized as fixed, partially floating, or floating systems. Floating systems, in particular, offer unique advantages for long-span bridges because of their remarkable spanning capacity and high flexibility [1], which significantly reduces seismic demands during near-fault earthquakes, providing exceptional seismic mitigation performance. However, the high flexibility of floating systems poses critical challenges under strong seismic events, as the absence of rigid pylon-deck connections often results in substantial deck displacements. This can lead to excessive deformation of the pylons or overstressing of their bases, ultimately threatening the structural integrity of bridges during seismic excitations [2–4].

In recent decades, extensive research focused on mitigating the excessive displacements in floating cable-stayed bridges under seismic loading. To enhance their seismic performance, various pylon-deck connection devices have been proposed. These devices can be broadly categorized into two types: (1) isolation devices, such as elastic bearings and friction pendulum bearings, which extend the fundamental periods of bridges to reduce seismic demands [5–7], and (2) energy dissipation devices, such as viscous dampers and metallic dampers, which absorb seismic energy through material yielding or thermal dissipation [8].

Among these, Fluid Viscous Dampers (FVDs) demonstrate superior performance in vibration control, owing to their exceptional energy dissipation capabilities [9–11]. Extensive studies have highlighted the effectiveness of FVDs in reducing vibration amplitudes and seismic responses [12–16]. For instance, Zhu et al. [4] reported that nonlinear FVDs can reduce longitudinal deck displacement by up to 79% and tower bending moment by 56% in a cable-stayed bridge under randomly generated earthquakes. Furthermore, FVDs enhanced seismic performance without significantly altering the inherent stiffness of the structure, making them an adaptable and efficient solution for seismic response mitigation in cable-stayed bridges [4,17,18].

In the seismic design of long-span bridges, precise tuning of FVD parameters, particularly the velocity exponent (α) and damping coefficient (C), is essential for effective displacement control [19]. Feng et al. [20] demonstrated that the velocity exponent significantly influences displacement control, while the damping coefficient primarily governs energy dissipation efficiency. Wu et al. [21] conducted a parametric analysis of α and C to optimize multiple longitudinal seismic responses—namely, top displacement, base shear, and base moment of pylons—for the Xigu Yellow River cable-stayed bridge, showcasing the effectiveness of FVD optimization in achieving balanced seismic performance. Similarly, Wen et al. [22] utilized a genetic algorithm with parallel computation for multi-objective parameter optimization, aiming to minimize repair costs and enhance the seismic resilience of a benchmark cable-stayed bridge. Despite these advances, optimizing FVD parameters remains a considerable challenge, requiring the careful trade-off of multiple, often competing objectives, such as displacement control, stress mitigation, and structural feasibility [21–23].

In recent years, various methods have been proposed to enhance the efficiency of FVD parameter optimization. For example, He et al. [19] used the surface fitting function to optimize FVD parameters in a cable-stayed bridge under high-intensity seismic conditions, achieving a substantial reduction in the seismic response. Similarly, Xu et al. [23] employed the response surface method to identify optimal FVD parameters and validated their effectiveness in mitigating seismic responses through experimental testing. Liu et al. [24] developed a multi-objective optimization function for a single-pylon cable-stayed bridge based on the energy dissipation theory, which provided an optimal FVD parameter combination. While these studies achieved improvements in optimization efficiency, they were constrained by a limited number of design variables and struggled to identify optimal solutions within complex multi-objective optimization scenarios.

For multi-objective optimization design, data-driven optimization approaches have gained attention in recent engineering research. For instance, Chen et al. [25] utilized FEM simulations integrated with a response surface optimization algorithm to optimize the design parameters of link beams for seismic design of piers with varying heights. Similarly, Baei and Terzic [26] employed parallel computing and the multi-objective particle swarm optimization algorithm (MOPSO) to optimize viscous damper parameters, enhancing the seismic performance of a moment frame. Additionally, Sun et al. [27] applied the Extreme

Gradient Boosting Tree (XGBoost) algorithm combined with the analytic hierarchy process to evaluate ultra-high-performance concrete, effectively balancing multiple design objectives.

In conventional multi-objective optimization methods, the evaluation of objectives mostly depends on computationally intensive FE simulations. These high computational demands often hinds the exploration of high-dimensional design spaces, reducing the likelihood of identifying globally optimal solutions. To overcome this challenge, researchers have proposed some computationally efficient surrogate models as alternatives to FE simulations. For instance, Guo et al. [28] used an ANN-based surrogate model to accurately predict the design parameters of triple friction pendulum bearings (TFPBs) for high-speed railway bridges and derived the design parameters based on response evaluation. Fang et al. [29] developed a multi-objective optimization framework for hybrid-braced structures by integrating five distinct machine learning (ML) prediction models with the Non-Dominated Sorting Genetic Algorithm II (NSGA-II) to enhance design efficiency and performance. Jiang et al. [30] utilized the NSGA-II algorithm to optimize magnetorheological damper designs with a focus on two objectives, achieving effective trade-offs between mechanical and electromagnetic performance. Yang et al. [31] proposed an NSGA-II-based approach to optimize the repair sequencing of cable-stayed bridges within a three-objective framework. Similarly, Tian et al. [32] applied an improved NSGA-II to perform multi-objective optimization of cable forces in arch bridges.

Table 1 summarizes some state-of-the-art multi-objective optimization applications. While these studies demonstrate that NSGA-II is a robust algorithm for addressing multi-objective optimization due to its ability to identify Pareto-optimal solutions and maintain solution diversity, several critical challenges persist in optimizing FVD parameters. Specifically: (1) existing surrogate models struggle to achieve the necessary balance between accuracy and generalizability, which is crucial for reliability optimization results; (2) multi-objective optimization algorithms like NSGA-II necessitate numerous iterations per generation, making computationally intensive FE simulations impractical under tight time constraints. Furthermore, current research has explored only a limited range of design variables and typically restricted optimization objectives to no more than three, thereby constraining the optimization process. This indicates that state-of-the-art surrogate models often compromise either the accuracy or the generalizability of the optimization outcomes. Consequently, despite its potential, the application of NSGA-II for FVD parameter optimization in cable-stayed bridges remains underexplored. Addressing these challenges calls for the development of a computationally efficient, multi-task surrogate model to facilitate accurate and practical multi-objective optimization of FVD parameters.

This study aims to validate a robust multi-objective optimization strategy for the seismic design of FVDs in cable-stayed bridges. Two strategies are compared. The first relies on the parametric analysis of a cable-stayed bridge equipped with a single equivalent FVD, where optimal parameters are identified through engineering expertise. The second strategy leverages a data-driven surrogate model to accurately approximate seismic responses and pairs it with the NSGA-II optimization algorithm. This strategy reduces the dependency of forward evaluations on computationally intensive FE simulations during the optimization process, making it highly effective for tackling complex optimization problems involving multiple design parameters and objectives.

This paper is organized as follows: Section 2 describes the development of the finite element model for the cable-stayed bridge and provides details on the seismic input. Section 3 is dedicated to the analysis and optimization of FVD parameters through detailed parametric analysis. Section 4 introduces a surrogate model-based approach to optimize the FVD parameters for the case study bridge, utilizing an ANN surrogate model in combination with the NSGA-II. Finally, the paper concludes with a comparative analysis of the two-parameter optimization strategies.

Table 1. Review of state-of-the-art algorithms for multi-objective optimization.

Literature	Year	Optimization Strategy	Application
Xu et al. [23]	2019	FE model, parametric analyses	Parameter optimization of FVDs in cable-stayed bridges
Chen et al. [26]	2022	FE model, response surface method	Two-objective seismic optimization of design parameters in double-column piers
Baei et al. [27]	2022	FE model, MOPSO algorithm	Multi-objective optimization of damper parameters for viscous dampers in moment frames
Guo et al. [29]	2024	ANN model, response evaluation	TFPB parameter design based on ANN inverse prediction model in simply-supported bridges
Fang et al. [30]	2022	Five ML models, NSGA-II	Multi-objective optimization of brace design parameters in structures with hybrid braces
Jiang et al. [31]	2022	Design equations, NSGA-II	Multi-objective optimization of design parameters for a magnetorheological damper
Yang et al. [32]	2024	FE model, NSGA-II	Multi-objective optimization of repair sequence in cable-stayed bridges
Tian et al. [33]	2024	FE model, NSGA-II	Multi-objective optimization of cable forces in arch bridges

2. Methodology

This study focuses on optimizing the parameters of FVDs installed on a cable-stayed bridge by employing two distinct optimization strategies. As illustrated in Figure 1, the overall research framework is structured as follows:

(1) Finite element (FE) model development:

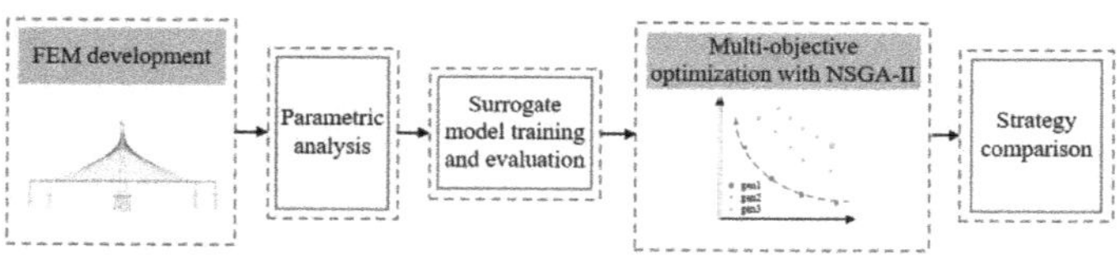

Figure 1. Optimization workflow of FVD damper parameters based on a surrogate model.

A detailed finite element model of the case study cable-stayed bridge is constructed. Representative ground motion records are selected as seismic inputs for the subsequent analysis and optimization tasks.

(2) Parametric analysis:

The seismic response of the case study bridge is evaluated under diverse combinations of FVD parameters. The relationships between damper parameters and seismic responses are analyzed to determine a practical range of design candidates.

(3) Surrogate model construction and evaluation:

A dataset is generated through nonlinear time-history analyses linking FVD parameters with seismic responses. Four machine learning models are trained and evaluated using metrics. The superior model (ANN) is selected to substitute computationally intensive FEM analyses during optimization.

(4) Multi-objective FVD optimization with NSGA-II:

The NSGA-II algorithm is utilized to solve a multi-objective optimization problem, aiming at three or four seismic performance objectives. Final optimal solutions are selected from the converged Pareto fronts.

(5) Comparison of optimization strategies:

The two optimization strategies, parametric analysis, and surrogate model-based op-timization, are compared in terms of their methodologies and the seismic responses of the bridge under the optimal parameters derived from each strategy.

2.1. Finite Element Model of the Case Study Cable-Stayed Bridge

The case study bridge is the Shijiazhuang Hutuo River Bridge, featuring a main span arrangement of 40 m + 150 m + 150 m + 40 m. It is a single-pylon cable-stayed bridge with uniquely twisted spatial stay cables symmetrically arranged on both sides of the pylon. The bridge deck comprises two separate steel box girders, each 29.5 m wide, connected by transverse girders spaced 8 m apart. The pylon is centrally located along the bridge's total longitudinal length and within the transverse girder spacing. The structural configuration and geometric dimensions are illustrated in Figure 2.

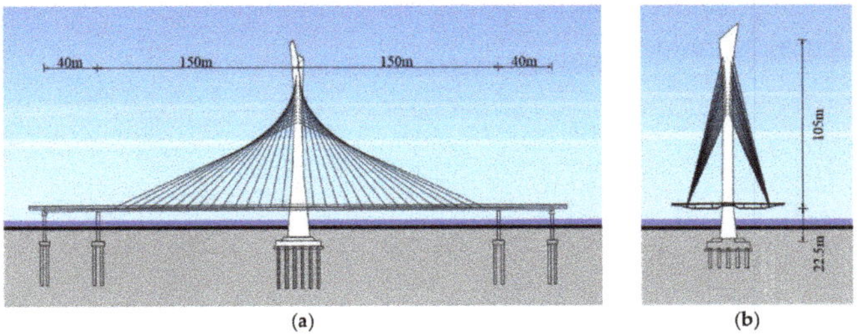

Figure 2. Hutuo River cable-stayed bridge: (**a**) longitudinal view; (**b**) transverse view.

This study develops a three-dimensional (3D) numerical model of the cable-stayed bridge using the OpenSees finite element framework (version 3.6.0) [33], as shown in Figure 3. The model represents the entire bridge structure, including two side spans and two middle spans, with 706 nodes and 707 elements. Specifically, the finite element configuration comprises 332 girder elements, 67 pylon elements, 62 cable elements, 204 rigid links, and 1 zero-length element. Elastic beam-column elements are used to model both the pylon and the longitudinal girders. Between the two longitudinal girders, there are forty transverse girder elements that connect to both longitudinal girders with eighty rigid links. The upper part of the pylon located above the deck is made of steel and is modeled with an elastic modulus of 2.06×10^{11} N/m^2, while the lower part of the pylon beneath the deck is made of concrete with an elastic modulus of 3.6×10^{10} N/m^2. The stay cables are represented by truss elements using the *steel02* material in OpenSees, characterized by an elastic modulus of 195 GPa and yield stress of 1.86 GPa. The 62 cables are connected to the pylon and the longitudinal girders with 124 rigid links on both ends. All the constitutive behaviors of these elements are illustrated in Figure 3.

The case study cable-stayed bridge is modeled as an ideal floating system. To control the displacements between the pylon and the longitudinal girders, an equivalent FVD is installed between the pylon and the girders, as illustrated in Figure 3. The equivalent FVD is based on the Maxwell viscoelastic model, which combines a linear spring with a nonlinear dashpot in series [4]. This model simplifies the analysis by reducing the design parameters to three: damper stiffness (K), damping coefficient (C), and velocity exponent (α), which are physically intuitive and easy to interpret. In the bridge's FE model, a single equivalent FVD is built with a zero-length element, and its mechanical behavior is governed by the Maxwell model. The piers beneath the side spans are modeled as roller supports that only permit longitudinal sliding.

After developing the FE model, the dynamic responses of the cable-stayed bridge equipped with FVDs were evaluated under various parameter combinations using nonlinear time-history analysis. To streamline the analysis, the pylon base was assumed to be fully fixed to the ground, thereby disregarding soil-structure interaction (SSI). Live load effects in both lateral and vertical directions were excluded from consideration in the dynamic analysis, focusing solely on the seismic performance of the bridge [34]. Additionally, cable sag effects in the stay cables are omitted in this study to simplify the analysis. While structural nonlinearity and SSI are known to significantly influence the seismic behavior of long-span bridges under extreme loading conditions [4,35], their inclusion would introduce increased model complexity, computational demands, and uncertainties related to soil properties. As these factors are beyond the scope of this study, they were excluded from focusing on optimizing the FVD parameters and evaluating key superstructure responses to seismic excitation. Furthermore, the adoption of linear assumptions ensures consistency in comparison with results from previous studies [36].

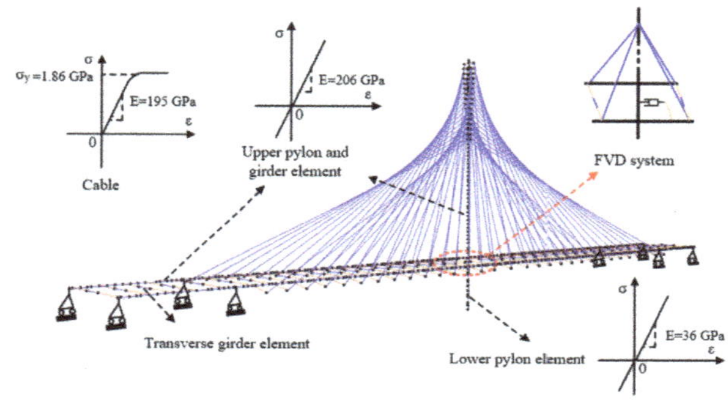

Figure 3. Finite element model of the case study cable-stayed bridge and the implementation of FVD system.

2.2. Construction of Surrogate Models

(1) Seismic performance evaluation indicators

The seismic performance evaluation indicators for the cable-stayed bridge in this study—longitudinal displacement of the deck, longitudinal top displacement of the pylon, and the base shear of the pylon—are key measures for understanding its behavior under seismic loads. Longitudinal deck displacement reflects the horizontal seismic response, deformation characteristics, and stiffness distribution of the bridge, highlighting the efficiency of the seismic design. The pylon's top displacement measures its deformation and structural stability, emphasizing the stiffness and load transmission of the pylon under earthquake actions. Base shear at the pylon's base provides insights into seismic force pathways and highlights the foundation's risk, particularly in understanding how forces are transmitted through the pylon to the substructure. Together, these indicators align with performance-based seismic design principles and provide a set of seismic performance indicators for a comprehensive evaluation of the cable-stayed bridge.

(2) Dataset construction of seismic responses

Based on the empirical design cases, the initial feasible variation ranges for the FVD parameters were determined, forming a three-dimensional design space. To efficiently explore this design space, the Latin Hypercube Sampling (LHS) method was employed to generate sample data due to its capacity to minimize parameter correlations, maintain statistical diversity, and reduce the required sample size. This approach ensured This approach ensured a comprehensive representation of the parameter distribution [37]. Here, each FVD, defined by a parameter combination (K, C, α), contains three design parameters. Using the LHS method, 1000 representative parameter combinations were generated (10 × 10 × 10). Nonlinear time-history analyses of the FVD-equipped bridge were then performed on the OpenSees platform for each parameter combination to evaluate seismic responses. The damping parameters served as inputs, while seismic response indicators—longitudinal deck displacement (y_d), pylon top displacement (y_t), pylon base shear (y_s), and mean bending moment energy (y_e)—constituted the outputs, constructing the 'parameters—responses' dataset.

(3) Establishment and evaluation of surrogate models

Four machine learning frameworks such as Decision Tree (DT), Extreme Gradient Boosting (XGBoost), Random Forest (RF), and Artificial Neural Network (ANN) were selected as surrogate models to predict the seismic responses from the bridge with various FVD parameters. To ensure robust generalization to unseen data, the 'parameters—responses' dataset was divided into training, validation, and testing subsets. For the DT, RF, and XGBoost models, critical hyperparameters, including the number of decision trees, maximum tree depth, learning rate, number of base estimators, and minimum samples per split, were systematically optimized using the Optuna framework (version 1.4.0) [38], ensuring improved model accuracy and robustness. Similarly, for the ANN model, essential hyperparameters such as the number of layers, neurons per layer, and learning rate were fine-tuned to maximize the network's learning efficiency and generalization capabilities.

Model performance evaluation was conducted using five-fold cross-validation to minimize variability and ensure robust assessment. Metrics such as the coefficient of determination (R^2), Root Mean Squared Error (RMSE), and Mean Absolute Error (MAE) were calculated to provide a comprehensive evaluation of prediction accuracy. Results from both the training and testing phases were assessed to detect and mitigate overfitting and underfitting risks. Ultimately, the surrogate model with the best predictive performance was selected and applied to optimize the FVD parameter design.

2.3. Multi-Objective Optimization Using the NSGA-II Algorithm

Under seismic loading, significant relative displacement between the pylon and the deck of a floating-type cable-stayed bridge requires effective mitigation strategies, such as installing FVDs between the pylon and the deck. This study focuses on optimizing three key seismic performance objectives in the longitudinal direction: pylon top displacement, longitudinal deck displacement, and pylon base shear. To achieve this, the Non-Dominated Sorting Genetic Algorithm II (NSGA-II)—a powerful multi-objective optimization algorithm based on an elitist strategy—is employed. NSGA-II offers several advantages, including low computational complexity, fast optimization speed, high accuracy, and diverse solution sets, making it well-suited for multi-objective optimization [39]. In this study, NSGA-II is used to derive the Pareto front, representing optimal trade-off solutions for the three seismic objectives. From the Pareto front, an optimal solution is selected to balance the engineering demands for longitudinal deck displacement, pylon top displacement, and pylon base shear. The results demonstrate the effectiveness of NSGA-II in optimizing FVD parameters to address complex seismic performance demands.

2.4. Selection of Ground Motions

Near-fault ground motions pose significant challenges to flexible bridges such as long-span cable-stayed bridges. These challenges arise from their pronounced velocity pulse effects and high ground velocity-to-acceleration ratios, which focus energy predominantly in the low-frequency range (0.1–1.0 Hz) [40]. Notably, the natural frequencies of cable-stayed bridges often fall within 0.1–0.5 Hz. This study targets a longitudinal floating-type cable-stayed bridge that is especially sensitive to near-fault motions. Under such seismic conditions, the floating configuration leads to prominent longitudinal deck displacements and horizontal displacements at the pylon tops.

To ensure the analysis accurately reflects the seismic performance of the case study cable-stayed bridge, near-fault earthquake records were carefully selected. Seven representative near-fault ground motion records were chosen from the Pacific Earthquake Engineering Research Center (PEER) database, encompassing a diverse range of tectonic environments to account for variability in seismic hazards. The key ground motion characteristics, summarized in Table 2, include parameters such as earthquake magnitude, the average shear wave velocity of the top 30 m of soil (Vs_30), peak ground acceleration (PGA), and peak ground velocity (PGV). The selected records have PGA values ranging from 0.28 g to 1.43 g.

Table 2. Selected near-fault ground motion records.

No.	Event	Year	Station	Magnitude	Vs_30 (m/s)	PGA (g)	PGV (cm/s)
1	Imperial Valley-06	1979	Chihuahua	6.53	242.05	0.28	30.5
2	Loma Prieta	1989	BRAN	6.93	476.54	0.64	55.9
3	Cape Mendocino	1992	Cape Mendocino	7.0	514	1.43	119.5
4	Northridge-01	1994	LA—Sepulveda VA	6.7	380	0.73	70.1
5	Kocaeli, Turkey	1999	Yarimca	7.5	297	0.31	73
6	Chi-Chi, Taiwan	1999	TCU084	7.6	553	1.16	115.1
7	Denali, Alaska	2002	TAPS Pump Sta. #10	7.9	553	0.33	126.4

According to the Chinese Guidelines for Seismic Design of Highway Bridges [41] and the seismic hazard report for the target bridge site, the design seismic level is defined by a PGA of 0.20 g and a characteristic period of 0.4 s. The site is classified as Type II within a seismic zone of 8 degrees defined by Chinese codes. To account for local hazard conditions, the selected seven ground motion records were scaled to a PGA of 0.20 g. This scaling aligns with the dynamic characteristics and seismic site conditions of the case study bridge. After amplitude adjustments, the scaled ground motions were utilized to develop the design response spectrum, as illustrated in Figure 4 [42]. Given the bridge's pronounced sensitivity to longitudinal displacements, the seismic analysis was conducted exclusively under longitudinal seismic excitations to capture its critical dynamic response.

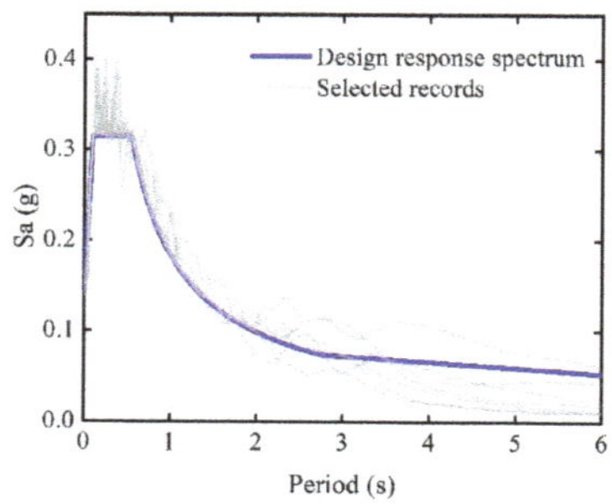

Figure 4. Acceleration spectra for the scaled ground motions.

3. Results and Discussion

3.1. Optimization of FVD Based on Parametric Analysis

3.1.1. Parametric Analyses of FVDs

In the seismic design of the case study cable-stayed bridge, an equivalent FVD is installed between the pylon and the girders. The design parameters of the equivalent FVD include damping coefficient (C), velocity exponent (α), and damper stiffness (K). For the sensitivity analysis of these parameters regarding structural seismic responses, an equally spaced sampling approach was utilized across their respective ranges. The damping coefficient (C) was set between 1000 and 10,000 kN/(m/s), with intervals of 1000 kN/(m/s). The velocity exponent (α) varied from 0.1 to 1.0, with intervals of 0.1, while the damper stiffness (K) ranged between 1×10^5 and 1×10^6 kN/m, with intervals of 1×10^5 kN/m. Using the identical design response spectrum, the seismic responses of the case study bridge equipped with FVDs were evaluated under various parameter combinations, with a focus on three critical performance metrics: longitudinal deck displacement, pylon top displacement, and pylon base shear.

Figures 5–7 illustrate the variations in the three seismic responses under the influence of different FVD parameters (K, C, and α). It can be observed that all three seismic responses are nearly insensitive to changes in damper stiffness. Specifically, as damper stiffness increases from 2×10^5 kN/m to 1×10^6 kN/m, the seismic responses exhibit a slight decrease. This observation aligns with previous studies [43] that indicate damper stiffness exerts a negligible influence on the seismic performance of bridges. Building on these findings, a damper stiffness (K) within the range of 8×10^5 to 1×10^6 kN/m is recommended.

In contrast, the velocity exponent and damping coefficient show a stronger correlation with these responses, consistent with findings from previous reports [4,44]. More specifically, both deck displacement and pylon top displacement decrease with an increase in C, but increase with a higher α. Conversely, the pylon base shear exhibits the opposite trend: it increases with increasing C and decreases with increasing α. Notably, when C is as low as 1000 kN/(m/s), the deck displacement and base shear increase significantly, while the pylon top displacement decreases remarkably, as illustrated in Figure 6. This suggests that the damping effect diminishes and becomes localized around the pylon-deck connection. Therefore, a damping coefficient of C = 1000 kN/(m/s) is excluded from the design range of damper parameters.

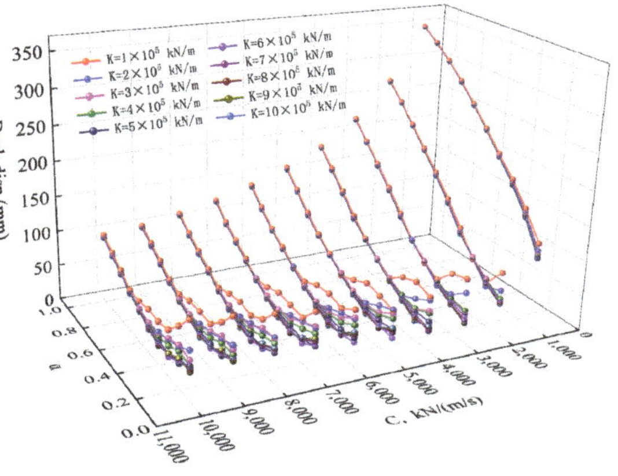

Figure 5. Longitudinal deck displacement as a function of C and α across different damper stiffnesses.

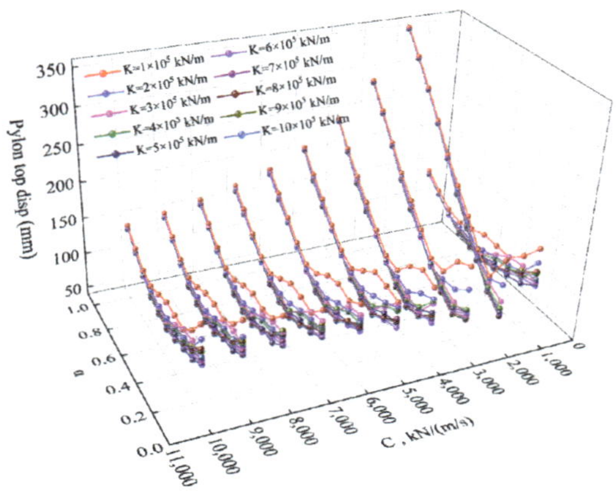

Figure 6. Pylon top displacement as a function of C and α across different damper stiffnesses.

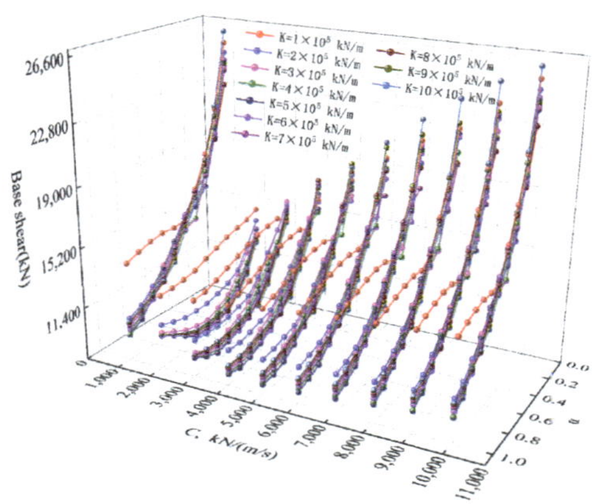

Figure 7. Pylon base shear as a function of C and α across different damper stiffnesses.

3.1.2. Optimal Design Spaces for Damper Parameters

To consider the importance of each response, all responses corresponding to each parameter combination are first normalized. In this study, Min-Max normalization is employed to scale the data to a uniform range of [0, 1]. This normalization addresses the issue of dimensional inconsistencies among the various responses [45] that are calculated in different units. Furthermore, this allows the adaptive weighting of these responses based on their relative importance, ensuring a balanced consideration in subsequent parametric analyses. Then, the summation of the normalized responses g (=$\Sigma\omega_i R_i$) is determined based on the normalized responses and expressed as Equation (1):

$$g = \frac{1}{n}\sum_{i=1}^{n}\omega_i R_i = \frac{1}{n}\sum_{i=1}^{n}\omega_i \frac{r_i - Min(r_i)}{Max(r_i) - Min(r_i)} \quad (1)$$

where R_i represents the i-th normalized response (i = 1, 2, …, n); n is the number of considered responses; ω_i represents the weight of the i-th response; r_i is the i-th original response; Min(r_i) and Max(r_i) denote the minimum and maximum values, respectively, of the i-th original responses across all data points.

This approach transforms the multi-objective optimization problem into a weighted sum of the multiple objectives [46]. By minimizing the single optimization function g, the overall optimization can be achieved. During parametric analysis, the corresponding g value was calculated for each parameter combination. The top ten smallest g values were selected, and the ten corresponding parameter combinations are presented in Table 3. From Table 3, it is observed that the velocity exponent falls in the range from 0.2 to 0.5, the damper stiffness from 8×10^5 to 1×10^6 kN/m, and the C from 3000 and 6000 kN/(m/s). These ranges are consistent with the results depicted in Figures 5–7.

Table 3. Optimal damping parameter combinations.

Velocity Exponent	Damper Stiffness (kN/m)	Damping Coefficient kN/(m/s)	Deck Disp. (mm)	Pylon Top Disp. (mm)	Pylon Base Shear (kN)	g
0.3	1×10^6	3000	54.7	92.3	10,847.8	0.276
0.2	1×10^6	3000	40.0	77.1	12,431.5	0.278
0.3	9×10^5	3000	54.4	91.6	10,971.7	0.280
0.2	9×10^5	3000	40.4	76.9	12,536.9	0.281
0.4	1×10^6	4000	55.9	93.6	10,858.8	0.283
0.3	1×10^6	4000	41.6	81.2	12,310.3	0.284
0.3	8×10^5	3000	54.5	91.8	11,125.0	0.286
0.4	1×10^6	5000	45.8	86.3	11,901.9	0.288
0.4	9×10^5	4000	55.9	93.1	11,011.6	0.288
0.5	1×10^6	6000	51.2	90.4	11,463.3	0.291

Figure 8 illustrates the three crucial responses within a multi-objective design space. In this space, a red point represents a response combination derived from the bridge equipped with a specific parameter combination. The blue points represent the ten selected responses that meet the minimization of the function g. It is observed that, in this case study, the optimal response combination can be clearly observed because the pylon top displacement and deck displacement are correlated with each other. From the blue points, the corresponding damper parameter combinations can be determined.

The three normalized responses, R_i, are mapped onto a two-dimensional space defined by C and α, both of which significantly influence the structural response, as shown in Figure 9. To achieve lower deck and pylon top displacements, the optimal parameter space lies in the upper-left region of the C-α space, where C takes a larger value while α is smaller. In contrast, to minimize the pylon base shear, the optimal parameter space shifts to the lower-right region of the C-α space, where C takes a smaller value and α is larger.

Figure 9d illustrates the preferred parameter space of the damper based on the sum of the three normalized responses, g, under the assumption of equal weighting ($\omega_1 = \omega_2 = \omega_3 = 1.0$). The blue region in the figure indicates the global optimal range for the combined responses, corresponding to a C range of 2000 to 7000 kN/(m/s) and an α range of 0.2 to 0.5. As shown in Figures 5–7, higher damping stiffness effectively suppresses structural responses. Notably, this parameter selection region is consistent with the optimal ranges suggested by Table 3.

This parametric analysis-driven optimization method is straightforward and allows for intuitive interpretation. However, it has several notable limitations. First, the parameter sampling is non-uniform and depends on empirically predefined samples, which may restrict the exploration of the design parameter space. Second, due to data constraints, an exhaustive analysis of the design parameter space is not feasible, making it challenging to pinpoint the precise optimal damper parameters. As a result, while this approach offers valuable preliminary guidance for parameter selection, further optimization is necessary to improve accuracy and ensure practical applicability.

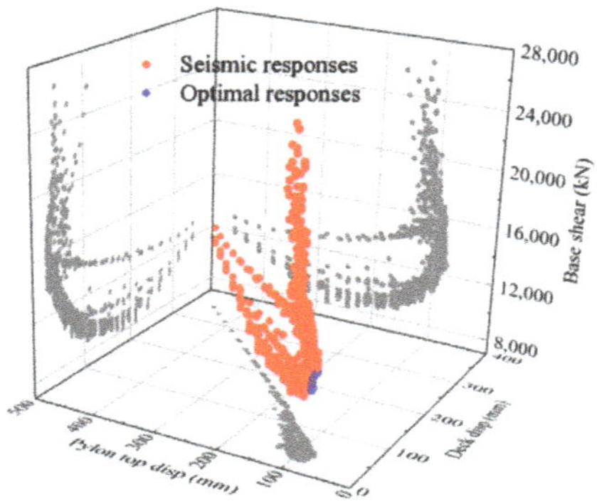

Figure 8. Multi-objective design space for three key structural responses.

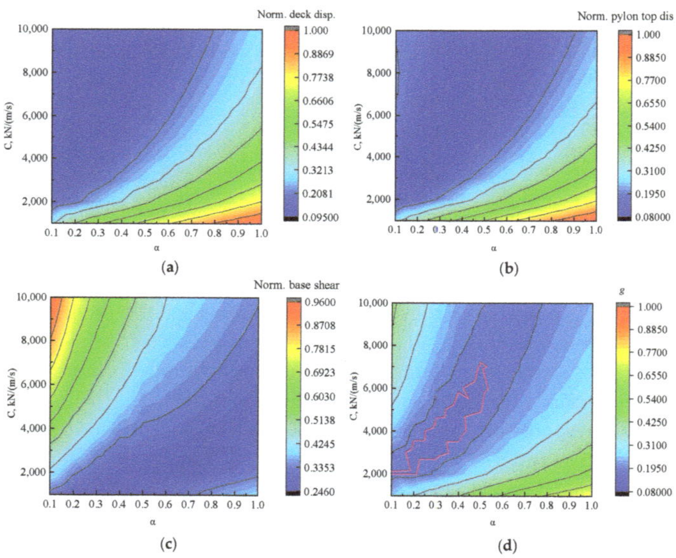

Figure 9. Projection of normalized responses onto the design space defined by C and α: (**a**) deck displacement, (**b**) pylon top displacement, (**c**) pylon base shear, and (**d**) the summation of the three normalized responses.

3.1.3. Seismic Mitigation of the Case Study Bridge

Figure 10 illustrates the time histories of three key structural responses of the case study bridge equipped with or without FVDs at the selected optimal parameter combination (α = 0.3, K = 1 $\times$ 10^6 kN/m, C = 3000 kN/(m/s)) under the design earthquake. In the figure, blue lines represent the time-history responses of the undamped bridge (without FVDs), while red lines depict those obtained with the optimized FVDs. The responses—longitudinal deck displacement, pylon top displacement, and pylon base shear—are reduced to 54.7 mm, 92.3 mm, and 10,847.8 kN, respectively, after the installation of FVDs. These results demonstrate that incorporating the optimally tuned FVDs significantly enhances the seismic performance of the case study bridge by effectively mitigating key seismic responses. Compared to the case study bridge without FVDs, the pylon top displacement experiences the largest reduction, decreasing by 88.3%. Similarly, the longitudinal deck displacement is reduced by 85.0%, and the pylon base shear decreases by 25.7%. These findings highlight the effectiveness of the optimized FVD parameters in controlling critical structural responses under seismic loading. Moreover, the observed effectiveness is consistent with findings from previous studies [4,47].

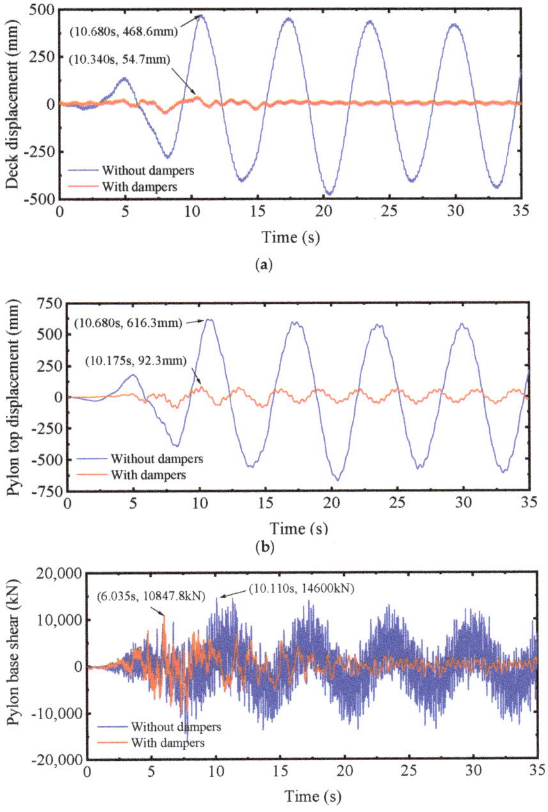

Figure 10. Time histories of (**a**) the longitudinal deck displacement, (**b**) the pylon top displacement, and (**c**) the pylon base shear for the case study bridge subjected to the design earthquake.

3.2. Surrogate Model-Based Parameter Optimization for FVDs

3.2.1. Dataset Generation

In this study, the seismic design of the case study bridge was optimized by calibrating the parameters of its FVDs. The selected FVD parameters included stiffness (K), damping coefficient (C), and velocity exponent (α). The design space for these three parameters aligns with the parameter sensitivity analysis strategy, with K ranging from 1×10^5 to 1×10^6 kN/m, C ranging from 1×10^3 to 1×10^4 kN/(m/s), and α ranging from 0.1 to 1.0. A total of 1000 parameter combinations were produced using the LHS method. For each combination, nonlinear time-history analyses were performed on FE models of the cable-stayed bridge equipped with FVDs, producing its responses under near-fault seismic excitations over a duration of 35 s. The key responses analyzed included longitudinal deck displacement, pylon top displacement, pylon base shear, and bending moments at all deck and pylon nodes. Additionally, the mean bending moment energy (MBME) of the bridge elements was calculated using Equation (2):

$$\text{MBME} = \frac{1}{m}\sum_{i=1}^{m} \frac{l_i}{4E_i I_i}\left({}^{L}M_i^2 + {}^{R}M_i^2\right) \tag{2}$$

where m represents the total number of structural elements; l_i, E_i, I_i, ${}^{L}M_i$, and ${}^{R}M_i$ denote the length, elastic modulus, bending moment of inertia, left-node bending moment, and right-node bending moment of the i-th element in the decks and pylons, respectively.

The peak values of deck displacement, pylon top displacement, and pylon base shear of the case study bridge undergoing a 35 s duration of ground motion are selected as the key seismic performance indicators. Additionally, MBME reflects mechanical rationality in terms of material usage and strength demand, whereas a lower MBME indicates reduced economic and strength requirements. Therefore, MBME can serve as an additional performance indicator, contributing to structural optimization in seismic design. A total of 1000 parameter-responses datasets were generated, with each dataset containing three critical FVD parameters and four output responses.

3.2.2. Training of Surrogate Models

Four types of machine learning models such as DT, XGBoost, RF, and ANN were constructed. All models utilized the identical dataset generated from the FE simulations. The dataset was randomly divided into three subsets: 70% for training, 15% for validation, and 15% for testing.

To fully leverage the potential of the DT, XGBoost, RF, and ANN models, we employed Optuna, an automated hyperparameter tuning framework, to fine-tune their hyperparameters. For the DT and XGBoost models, the respective search spaces for hyperparameter tuning are detailed in Tables 4 and 5. Using RMSE and MAE as evaluation metrics, the Optuna framework was employed to automatically optimize critical hyperparameters, such as the maximum depth (max_depth), the minimum number of samples required to split a node (min_samples_split), and the minimum number of samples required at a leaf node (min_samples_leaf) for the DT model. Similarly, key hyperparameters for the XGBoost model, including the number of estimators (n_estimators), the number of layers (n_layers), learning rate, subsampling ratio (subsample), and column subsampling by tree (colsample_bytree), were optimized. The finalized optimal hyperparameters are presented in Tables 4 and 5.

Table 4. Optimal DT hyperparameters and search ranges.

Hyperparameter	Optimal Value	Search Range
max_depth	17	1–20
min_samples_split	5	1–10
min_samples_leaf	2	1–10

Table 5. Optimal XGBoost hyperparameters and search ranges.

Hyperparameter	Optimal Value	Search Range
n_estimators	85	10–200
n_layers	6	3–10
learning_rate	0.1	0.01–0.1
subsample	0.9	0.6–1
colsample_bytree	0.9	0.6–1

For the RF model, we defined a search space for the hyperparameters (see Table 6) and evaluated performance using RMSD and MAE metrics. Optuna utilizes a Bayesian optimization algorithm implemented via the Tree-structured Parzen Estimator (TPE) to optimize key hyperparameters, such as the number of decision trees (n_estimators), maximum tree depth (max_depth), and the minimum number of samples required to split a node (min_samples_split), among others. This Bayesian approach effectively balances exploration and exploitation by constructing a probabilistic model of the objective function, enabling the algorithm to converge toward optimal hyperparameter settings within a limited number of iterations. The optimal hyperparameters identified for the RF model are presented in Table 6.

Table 6. Optimal RF hyperparameters and search ranges.

Hyperparameter	Optimal Value	Search Range
n_estimators	34	10–100
max_depth	5	3–10
min_samples_split	3	1–10
min_samples_leaf	1	1–10
bootstrap	True	—

Similarly, for the ANN model, the hyperparameter search space was defined to include the number of layers (n_layers), the number of neurons per layer (hidden_layer_size), and the learning rate, as outlined in Table 7. Using RMSE and MAE as evaluation metrics, Optuna was employed to automate the optimization of these hyperparameters. The resulting optimized hyperparameters for the ANN model are summarized in Table 7. The proposed ANN model, as depicted in Figure 11, is designed with three input features and four output features. Its architecture comprises four hidden layers with 16, 32, 64, and 32 neurons, respectively. Before training, input data were preprocessed using Min-Max normalization. The ReLU activation function was employed across all hidden layers to ensure nonlinear transformations. For training, the Adam optimizer was utilized with a learning rate of 0.008 and a maximum of 3000 iterations, enabling efficient convergence of the model.

Table 7. Optimal ANN hyperparameters and search ranges.

Hyperparameter	Optimal Value	Search Range
hidden_layer_size	16, 32, 64, 32	10–100
n_layers	4	2–5
activation	ReLU	—
solver	Adam	—
α	0.0001	0.1–0.00001
learning_rate	0.008	—

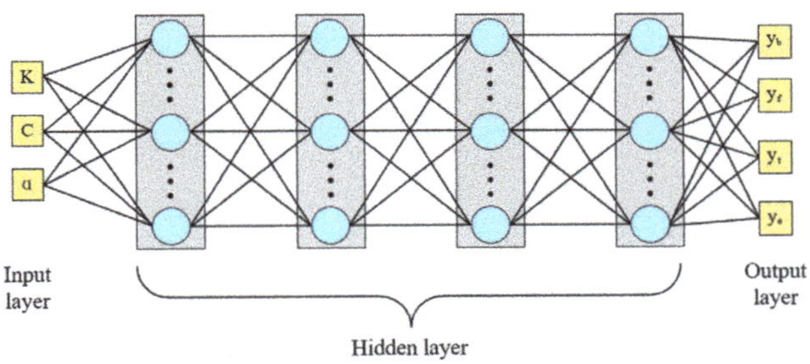

Figure 11. Architecture of the ANN prediction model.

3.2.3. Loss Function

To enhance the performance of the ANN model, this study employs the Mean Squared Error (MSE) as the loss function for each task. Specifically, the ANN model defines three separate loss components for the primary objectives: longitudinal deck displacement (L_1), pylon top displacement (L_2), and pylon base shear (L_3). Given that these three tasks are treated as equally important, their loss function weights are set identically ($\lambda_1 = \lambda_2 = \lambda_3$). In addition, an auxiliary task is incorporated to predict the mean bending moment energy, with its corresponding loss component (L_4). This auxiliary loss not only quantifies a key structural response but also serves as a regularization term, effectively balancing the contributions of the individual tasks during training.

To ensure consistency in magnitude across task-specific losses, the raw data were normalized prior to training. Here, we employed a weighted aggregation method to combine multiple losses into a single total loss, defined as $L_t = \sum\lambda_i L_i$ (i = 1, 2, 3, 4), where λ_i and L_i represent the weight and loss of the i-th task, respectively. The weighting scheme was subject to the constraints: $\lambda_1 = \lambda_2 = \lambda_3$ and $\lambda_1 + \lambda_2 + \lambda_3 + \lambda_4 = 4$. Under these constraints, we investigated the evolution of the total losses for various weight combinations. Figure 12 illustrates the evolution of the total training and validation loss over iterations for a specific weighting scenario (λ_1:λ_2:λ_3:λ_4 = 1:1:1:1). The figure shows that the total losses decrease rapidly at the beginning, converge gradually as training progresses, and stabilize after reaching an inflection point.

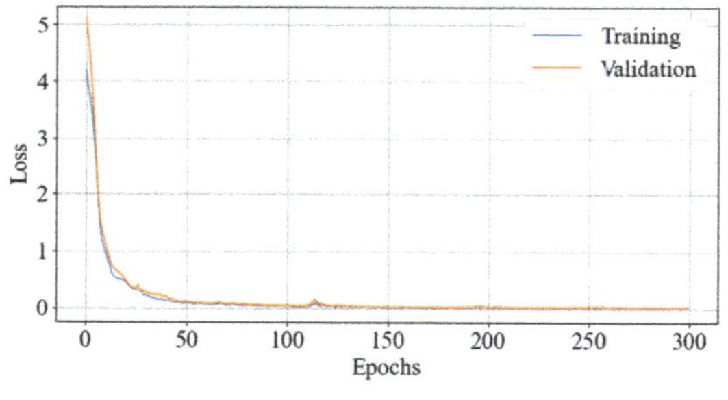

Figure 12. The total losses of the ANN model during training and validation processes.

Table 8 summarizes the key characteristics of the loss variation curves for different weight combinations. The results reveal that the optimal configuration is achieved with a weight ratio of λ_1:λ_2:λ_3:λ_4 = 1:1:1:1. In this setup, the model exhibits the most rapid convergence, marked by an inflection point at approximately 50 epochs. Furthermore, this combination achieves the lowest total training and validation losses, with a training loss of 0.012 and validation loss of 0.015, outperforming all other tested combinations. This indicates that aligning the weight of the MBME loss with those of the structural response losses significantly enhances the convergence rate of the ANN model.

Table 8. Characteristics of the total loss variation curves for different weight combinations.

λ_1:λ_2:λ_3:λ_4	Inflection (Epochs)	Total Training Loss	Total Validation Loss
1.3:1.3:1.3:0.1	100	0.020	0.031
1:1:1:1	**50**	**0.012**	**0.015**
2/3:2/3:2/3:1	100	0.015	0.019
1/3:1/3:1/3:3	100	0.020	0.025
0.1:0.1:0.1:3.7	100	0.015	0.020

3.2.4. Evaluation of Surrogate Models

To evaluate the predictive accuracy and generalization capability of the models, three performance metrics were employed: RMSE, MAE, and R^2. The formulas for these metrics are provided below as Equations (3)–(5):

$$\mathrm{RMSE} = \sqrt{\frac{1}{n}\sum_{i=1}^{n}\left(y_i - \hat{y}_i\right)^2} \tag{3}$$

$$\mathrm{MAE} = \frac{1}{n}\sum_{i=1}^{n}\left|y_i - \hat{y}_i\right| \tag{4}$$

$$R^2 = 1 - \frac{\sum_{i=1}^{n}\left(y_i - \hat{y}_i\right)^2}{\sum_{i=1}^{n}\left(y_i - \overline{y}\right)^2} \tag{5}$$

In these equations, y_i represents the true value of the i-th output, $\hat{y}_i$ is the corresponding predicted value, n is the total number of data points, and $\bar{y}$ denotes the mean of the true values. RMSE and MAE are essential metrics for assessing model error, with lower values reflecting better predictive performance. In contrast, R^2 measures the proportion of variance in the true values that is explained by the predicted values, with values closer to 1.0 indicating a better model fit.

Figure 13 represents the prediction accuracies of four output features for the RF and ANN surrogate models. The horizontal axis represents the ground truth values from FE simulations, while the vertical axis depicts the values predicted by the surrogate models. Data points for both models align with the y = x line, indicating high predictive accuracy.

As summarized in Table 9, all models achieve high prediction accuracy on multiple tasks. Notably, the ANN model outperforms the DT, XGBoost, and RF models, achieving high accuracies above 98% across all tasks. Furthermore, the ANN model exhibits considerably lower RMSE and MAE values compared to the DT, XGBoost, and RF models. These findings affirm the ANN model's precision and validity as the surrogate model for subsequent parameter optimization.

To evaluate the model's predictive performance on unseen data, we employed a five-fold cross-validation procedure. Specifically, the training dataset was randomly segmented into five equal folds; one served as the validation set while the other four were used for training. We computed the evaluation metrics such as MAE, RMSE, and R^2 across all folds to ensure a robust assessment of the model's performance. Figures 14–17 display the ANN model's performance for four prediction tasks: longitudinal deck displacement, pylon top displacement, pylon base shear, and mean bending moment energy, respectively. Radar charts indicate that both RMSE and MAE values remain consistently low across all folds, while the average R^2 value is approximately 0.99 with a standard deviation of ± 0.01. These results demonstrate that the ANN model exhibits excellent generalization capabilities, confirming its suitability for parameter optimization tasks.

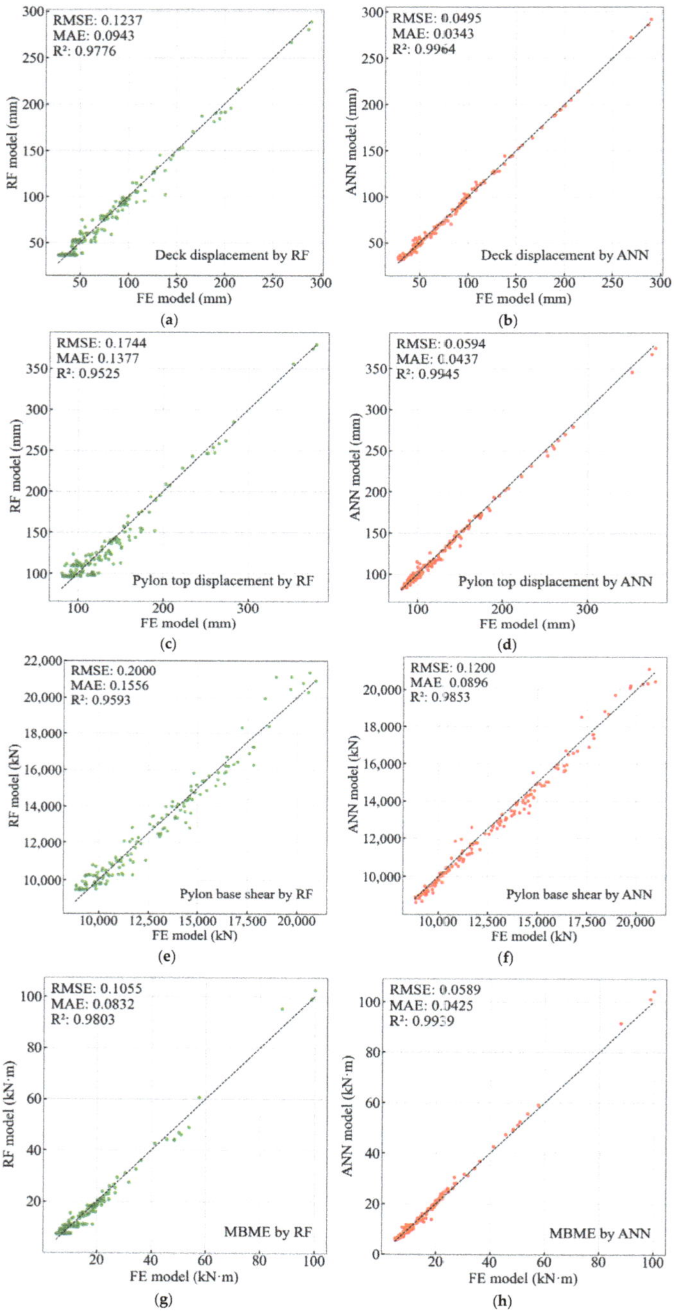

Figure 13. Comparison of prediction accuracy of (**a**,**b**) deck displacements, (**c**,**d**) pylon top displacement, (**e**,**f**) pylon base shear, and (**g**,**h**) MBME using RF and ANN models.

Table 9. Performance evaluation metrics of the DT, XGBoost, RF, and ANN surrogate models.

Model	Metrics	Deck Disp.	Pylon Top Disp.	Pylon Base Shear	MBME
DT	RMSE	0.140	0.161	0.234	0.105
	R^2	0.971	0.960	0.944	0.981
	MAE	0.094	0.106	0.174	0.067
XGBoost	RMSE	0.187	0.233	0.228	0.211
	R^2	0.948	0.915	0.947	0.922
	MAE	0.125	0.157	0.168	0.121
RF	RMSE	0.148	0.187	0.206	0.153
	R^2	0.979	0.967	0.951	0.979
	MAE	0.112	0.146	0.159	0.153
ANN	RMSE	0.060	0.069	0.124	0.058
	R^2	**0.995**	**0.993**	**0.984**	**0.994**
	MAE	0.044	0.047	0.080	0.039

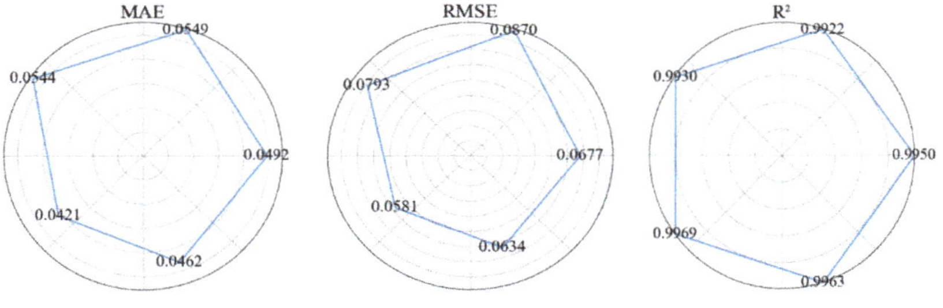

Figure 14. Evaluation metrics of five-fold cross-validation for longitudinal deck displacement.

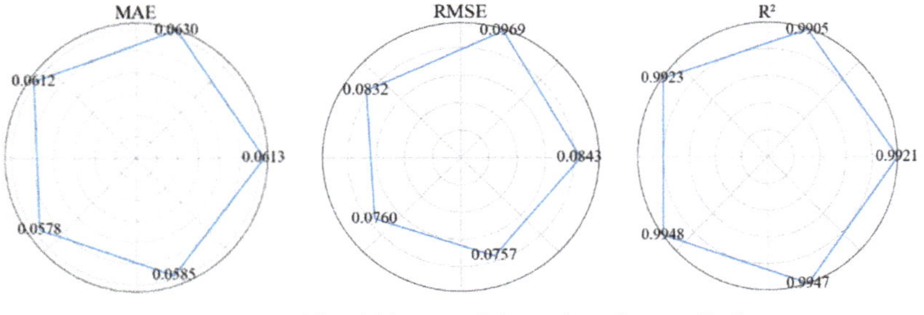

Figure 15. Evaluation metrics of five-fold cross-validation for pylon top displacement.

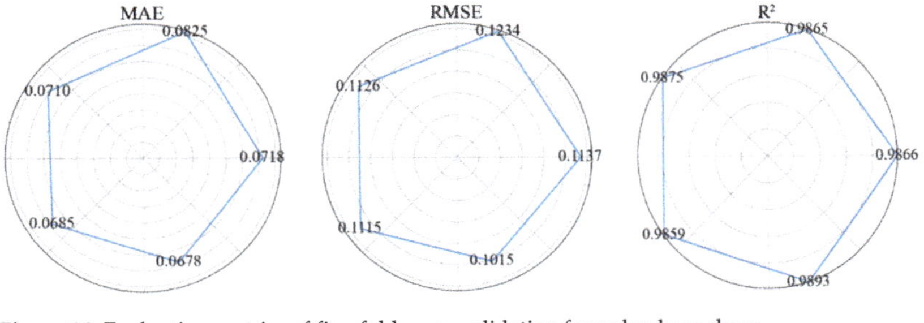

Figure 16. Evaluation metrics of five-fold cross-validation for pylon base shear.

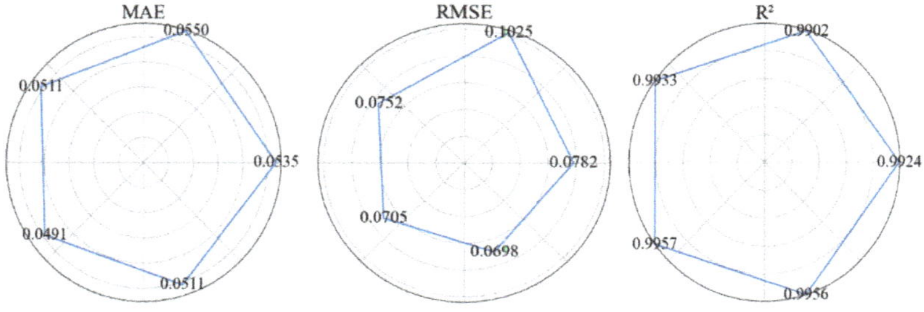

Figure 17. Evaluation metrics of five-fold cross-validation for mean bending moment energy.

3.2.5. Optimization of Damping Parameters Based on the ANN Model

In this study, the NSGA-II algorithm was employed to perform multi-objective optimization of FVD parameters. A population size of 200 was chosen to ensure adequate diversity and thorough exploration of the design space. For each generation, the structural responses for 200 distinct combinations of FVD parameters were rapidly predicted using a trained ANN surrogate model, which maintained high accuracy in predicting the targeted three or four objectives. After prediction, the outcomes were preliminarily evaluated, and candidate solutions underwent crossover and mutation operations. The crossover probability was set to 0.8 and the mutation probability to 0.1. These settings were chosen to balance the convergence rate with the preservation of solution diversity. Over 150 generations, the NSGA-II algorithm executed a total of 30,000 forward evaluations via the ANN surrogate model. This surrogate model-based approach effectively integrates the ANN with the NSGA-II algorithm, yielding significant computational efficiency improvements compared to conventional FE model-based methods, while preserving prediction accuracy.

Figure 18 illustrates the convergence trends of the summed mean normalized responses across generations for both the three-objective and four-objective optimization scenarios. The three-objective optimization focuses on minimizing: (1) pylon top displacement, (2) longitudinal deck displacement, and (3) pylon base shear, while the four-objective optimization introduces an additional goal—minimizing (4) mean bending moment energy. For both strategies, the normalized responses are averaged and aggregated, with rank and diversity assessed at each generation using ranking and crowding distance metrics. Subsequently, crossover and mutation operators are applied to drive the population toward improved Pareto-front solutions. Notably, in both scenarios, the summed mean normalized responses decrease rapidly during the initial stages, followed by a gradual increase before finally stabilizing at a specific value. These trends demonstrate the algorithm's effectiveness in achieving convergence within the design space.

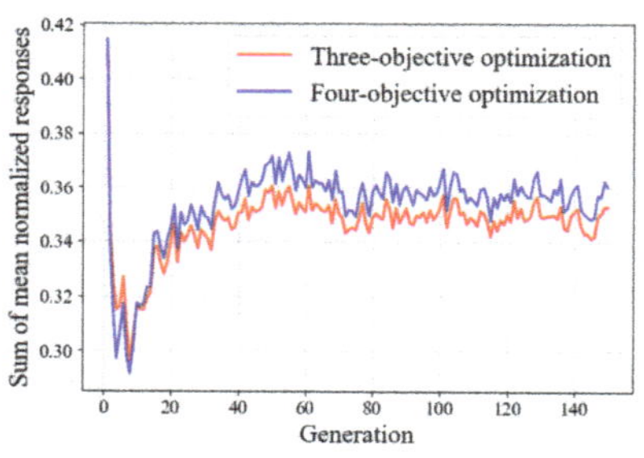

Figure 18. Convergence trends of the sums of mean normalized responses over generations for three-objective and four-objective optimizations.

Specifically, the three-objective optimization process can be visualized within a three-dimensional (3D) response space. Figure 19 demonstrates the evolution of the Pareto front across the 1st, 50th, and 150th generations for the three-objective optimization. Initially, the ANN surrogate model generated a broad 3D Pareto surface from 200 distinct parameter combinations, capturing an extensive design space. As the generation progressed, the Pareto surface gradually converged toward optimized solution sets, achieving convergence by the 150th generation.

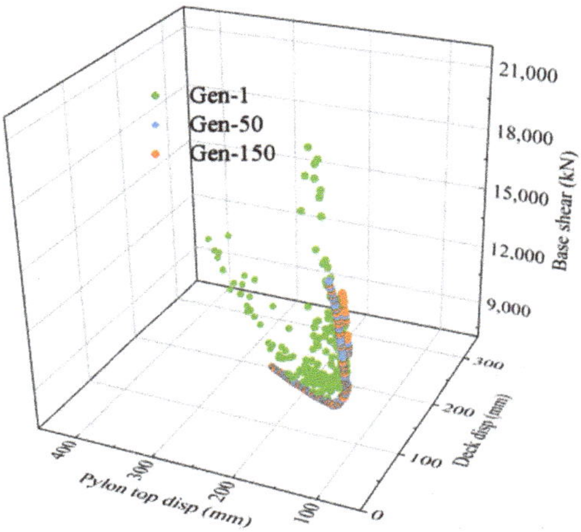

Figure 19. Evolution of Pareto fronts during the 3-objective optimization process.

Within the 3D response space, the Pareto front clearly illustrates the trade-offs among the three objectives and the evolution of responses across successive generations. A strong correlation is observed between pylon top displacement and deck displacement, alongside a clear trade-off between displacement responses and base shear. The optimization process demonstrates rapid convergence within the first 50 generations, after which the solutions become more densely distributed, reflecting a gradual convergence of the Pareto front.

Upon convergence of the Pareto front, the weighted aggregation method was applied to assess and rank the parameter combinations. This approach identified the optimal FVD parameters as K = 928,000 kN/m, C = 3938 kN/(m/s), and $\alpha = 0.32$. These parameters were subsequently incorporated into the FE model of the case study bridge for nonlinear time-history analyses. The FE simulation yielded a pylon top displacement of 45.2 mm, a deck displacement of 84.8 mm, and a base shear of 11,973.1 kN. Table 10 summarizes the key responses from both the FE simulations and the surrogate model predictions, revealing errors of up to 7.5% for deck displacement, 4.9% for pylon top displacement, and 3.0% for base shear. This indicates that the surrogate model provides predictions closely aligned with the FE simulations, even for FVD parameter combinations unseen in the training dataset.

Table 10. Performance metrics from the FE simulations and the surrogate model.

Performance Metric	ANN Model	FE Model	Error
Deck disp. (mm)	41.8	45.2	7.5%
Pylon top disp. (mm)	80.6	84.8	4.9%
Pylon base shear (kN)	11,611.3	11,973.1	3.0%

Table 11 summarizes the optimized FVD parameters obtained from three optimization strategies: parametric analysis-driven, ANN-based three-objective, and ANN-based four-objective optimization strategy. For each optimized parameter combination, four seismic responses of the case study bridge were computed via nonlinear time-history analyses. It can be noted that all seismic responses are significantly reduced when the optimized FVD is installed. Among the three optimization strategies, notably improved responses are highlighted in bold, as presented in Table 11.

Table 11. Comparison between three different optimization strategies.

Parameters/Responses	Without FVDs	Parametric Analysis-Driven Optimization	ANN-Based 3-Objective Optimization	ANN-Based 4-Objective Optimization
K (kN/m)	-	1,000,000	928,000	895,000
C (kN·s/m)	-	3000	3938	3947
α	-	0.3	0.32	0.33
Deck disp. (mm)	486.6	54.7	**45.2**	**47.4**
Pylon top disp. (mm)	613.3	92.3	**84.8**	**85.3**
Pylon base shear (kN)	14,600.0	**10,847.8**	11,973.1	**11,884.9**
MBME (kN·m)	121.2	10.1	8.9	**7.7**
Time consuming (min)	-	1500.5	1500.5 + 9.0 + 0.8	1500.5 + 9.0 + 0.8

In the parametric analysis-driven optimization strategy, the pylon base shear is effectively controlled at 10,847.8 kN, the lowest among the three strategies considered. Comparatively, the three-objective optimization strategy achieves a greater reduction in both deck displacement (45.2 mm) and pylon top displacement (84.8 mm), with these values being the smallest among the three strategies. However, this improvement comes with a slight increase in pylon base shear. On the other hand, the four-objective optimization strategy demonstrates a more balanced trade-off across the four responses. While displacement responses exhibit a minor increase (deck displacement: 47.4 mm; pylon top displacement: 85.3 mm), the base shear is notably reduced to 11,884.9 kN compared to the three-objective strategy.

Furthermore, the fourth response (MBME) was analyzed across all three strategies to assess the economic and mechanical rationality of the case study bridge. The smallest MBME value (7.7 kN·m) was achieved in the four-objective strategy, where the MBME was explicitly targeted during the NSGA-II optimization process. Interestingly, the three-objective strategy yielded a relatively low MBME value of 8.9 kN·m, even though it was not directly optimized during the process. This outcome demonstrates the advantage of incorporating the MBME response into the training of the ANN surrogate model, where it served not only as a predicted response but also as a loss penalty with clear physical significance [48]. In comparison, the parametric-analysis-driven strategy produced a higher MBME value of 10.1 kN·m.

Additionally, we compared the computational time required for the three strategies. For consistency, all strategies utilized an identical dataset, with data generation consuming 1500.5 min on a CPU-based computer (Intel i5, 2.90 GHz). Notably, this data generation time can be significantly reduced through parallel computation techniques. The training of the ANN surrogate model required only 9.0 min, while the optimization process itself was completed in approximately 0.8 min for 150 generations. These results demonstrate the efficiency and robustness of the ANN-based multi-objective optimization strategies in both training and optimization phases.

4. Conclusions

This study investigated two methods for optimizing FVD parameters in a case study cable-stayed bridge: a parametric analysis-driven strategy and a surrogate model-based strategy. Key conclusions are:

1. The parametric analysis-driven optimization strategy provides preliminary guidance for selecting optimal FVD parameters. Although this strategy is straightforward and interpretable, it relies heavily on expertise, restricting its ability to explore large design space and its generalization to multi-objective optimization problems.
2. The integration of an ANN surrogate model with the NSGA-II algorithm demonstrated that the ANN-based multi-objective optimization strategy can not only predict multiple seismic responses accurately, but also reduce computational cost during the multi-objective optimization process. This strategy enables efficient exploration of high-dimensional parameter spaces and yields more effective and generalizable solutions.
3. The ANN-based optimization strategy outperformed the parametric analysis-driven strategy by incorporating the MBME loss during surrogate model training. This enhancement improved the fidelity and generalization of the ANN model, demonstrating its potential for addressing high-dimensional, multi-objective optimization challenges.

While this study highlights the feasibility and advantages of the proposed optimization strategies for cable-stayed bridges with FVDs, it is based on several simplifying assumptions. For instance, the structural system is assumed to be linear, excluding considerations of soil-structure interaction and the sag of stay cables. To enhance practical applicability, future research should focus on improving the generalization capabilities of surrogate models. This can be achieved by: (1) expanding training datasets to include additional constraints and diverse parameters, such as varying loading conditions and seismic scenarios; and (2) incorporating nonlinear effects, including soil-structure interaction and cable sagging, to enhance model fidelity in real-world scenarios. These advancements would improve the adaptability of the proposed strategy, enabling it to address more complex and higher-dimensional engineering challenges in engineering practice.

Author Contributions: Q.L.: conceptualization, writing, review and editing, funding acquisition; Z.L.: investigation, software, writing—original draft preparation; J.Z.: investigation, methodology, writing—review and editing; Y.L.: software, methodology; S.Z.: writing—review, and editing, validation, funding acquisition; X.W.: writing—review, and editing, project administration. All authors have read and agreed to the published version of the manuscript.

Funding: This work was supported by the Open Foundation of State Key Laboratory for Strength and Vibration of Mechanical Structure under Grant SV2018-KF-35, the Start-Up Foundation of XUST under Grant 2018YQ2-05, and the Technology Innovation Foundation of CCCC under Grant KCJJ2023-32-07.

Data Availability Statement: Some data or models used during the study are available from the corresponding author by request.

Conflicts of Interest: Shimin Zhu and Xing Wu were employed by CCCC First Highway Consultants Co., Ltd. The remaining authors declare that the research was conducted in the absence of any commercial or financial relationships that could be perceived as a potential conflict of interest.

References

1. Xu, L.; Zhang, H.; Gao, J.; Zhang, C. Longitudinal Seismic Responses of a Cable-Stayed Bridge Based on Shaking Table Tests of a Half-Bridge Scale Model. *Adv. Struct. Eng.* **2019**, *22*, 81–93. [CrossRef]
2. Shen, Y.; Li, Y.; Xu, W.; Li, J. Evaluation of Seismic-Induced Impact Interaction between a Cable-Stayed Bridge and Its Approach Spans Using a Simplified Analysis Model. *J. Earthq. Eng.* **2022**, *26*, 3343–3363. [CrossRef]

3. Xie, W.; Sun, L. Transverse Seismic Response and Failure Mode of Towers of a Cable-Stayed Bridge Full-Model: Tests and Simulations. *Eng. Fail. Anal.* **2021**, *122*, 105224. [CrossRef]
4. Zhu, J.; Zhang, W.; Zheng, K.F.; Li, H.G. Seismic Design of a Long-Span Cable-Stayed Bridge with Fluid Viscous Dampers. *Pract. Period. Struct. Des. Constr.* **2016**, *21*, 04015006. [CrossRef]
5. Ali, H.M.; Abdel-Ghaffar, A.M. Seismic Energy Dissipation for Cable-stayed Bridges Using Passive Devices. *Earthq. Eng. Struct. Dyn.* **1994**, *23*, 877–893. [CrossRef]
6. Guo, W.; Zeng, C.; Xie, X.; Bu, D. Pseudodynamic Hybrid Simulation of High-Speed Railway Bridge-Track System with Rotational Friction Damper. *Int. J. Struct. Stab. Dyn.* **2020**, *20*, 2040014. [CrossRef]
7. Leblouba, M. Selection of Seismic Isolation System Parameters for the Near-Optimal Design of Structures. *Sci. Rep.* **2022**, *12*, 14734. [CrossRef]
8. De Domenico, D.; Ricciardi, G.; Takewaki, I. Design Strategies of Viscous Dampers for Seismic Protection of Building Structures: A Review. *Soil Dyn. Earthq. Eng.* **2019**, *118*, 144–165. [CrossRef]
9. Soneji, B.B.; Jangid, R.S. Passive Hybrid Systems for Earthquake Protection of Cable-Stayed Bridge. *Eng. Struct.* **2007**, *29*, 57–70. [CrossRef]
10. Calvi, G.M.; Sullivan, T.J.; Villani, A. Conceptual Seismic Design of Cable-Stayed Bridges. *J. Earthq. Eng.* **2010**, *14*, 1139–1171. [CrossRef]
11. Camara, A.; Astiz, M.A. Analysis and Control of Cable-Stayed Bridges Subject to Seismic Action. *Struct. Eng. Int.* **2014**, *24*, 27–36. [CrossRef]
12. Martínez-Rodrigo, M.D.; Filiatrault, A. A Case Study on the Application of Passive Control and Seismic Isolation Techniques to Cable-Stayed Bridges: A Comparative Investigation through Non-Linear Dynamic Analyses. *Eng. Struct.* **2015**, *99*, 232–252. [CrossRef]
13. Ruangrassamee, A.; Kawashima, K. Seismic Response Control of a Cable-Stayed Bridge by Variable Dampers. *J. Earthq. Eng.* **2006**, *10*, 153–165. [CrossRef]
14. Vader, T.S.; McDaniel, C.C. Influence of Dampers on Seismic Response of Cable-Supported Bridge Towers. *J. Bridge Eng.* **2007**, *12*, 373–379. [CrossRef]
15. Zhang, X.-J.; Zhao, C.-Y.; Guo, J. Investigation of Seismic Performance of Super Long-Span Cable-Stayed Bridges. *Earthq. Struct.* **2018**, *14*, 493–503.
16. Chang, C.-M.; Loh, C.-H. Seismic Response Control of Cable-Stayed Bridge Using Different Control Strategies. *J. Earthq. Eng.* **2006**, *10*, 481–508. [CrossRef]
17. Ali, H.E.M.; Abdel-Ghaffar, A.M. Seismic Passive Control of Cable-Stayed Bridges. *Shock. Vib.* **1995**, *2*, 259–272. [CrossRef]
18. Zoccolini, L.; Bruschi, E.; Pettorruso, C.; Rossi, D.; Quaglini, V. Fluid Viscous Dampers for Seismic Protection of Bridges: A State of the Art. *Procedia Struct. Integr.* **2024**, *62*, 669–676. [CrossRef]
19. He, X.; Yang, Y.; Xiao, X.; Deng, Y. Research on Fluid Viscous Damper Parameters of Cable-Stayed Bridge in Northwest China. *Shock. Vib.* **2017**, *2017*, 4532325. [CrossRef]
20. Feng, D.; Wang, J. Seismic Control of a Self-Anchored Suspension Bridge Using Fluid Viscous Dampers. *Int. J. Struct. Stab. Dyn.* **2021**, *21*, 2150025. [CrossRef]
21. Wu, C.; He, X.; He, L.; Zhao, X.; Wang, Y.; Li, C.; Yang, Y.; Zhang, X. Improving Cable-Stayed Bridge Longitudinal Aseismic Capability via Fluid Viscous Damper Parametric Optimization and Experimental Investigation. *Structures* **2023**, *57*, 105199. [CrossRef]
22. Wen, J.; Han, Q.; Xie, Y.; Du, X.; Zhang, J. Performance-Based Seismic Design and Optimization of Damper Devices for Cable-Stayed Bridge. *Eng. Struct.* **2021**, *237*, 112043. [CrossRef]
23. Xu, L.; Bi, K.; Gao, J.-F.; Xu, Y.; Zhang, C. Analysis on Parameter Optimization of Dampers of Long-Span Double-Tower Cable-Stayed Bridges. *Struct. Infrastruct. Eng.* **2020**, *16*, 1286–1301. [CrossRef]
24. Liu, Q.; Geng, J.; Wang, Y.; Ge, Y.; Bao, Q. Multi-Parameter Damper Optimization of Cable-Stayed Bridge Considering Energy Dissipation. *J. Phys. Conf. Ser.* **2023**, *2541*, 012002. [CrossRef]
25. Chen, X.; Spencer, B.F.; Li, J.; Guan, Z.; Pang, Y. Optimization of Distribution Patterns of Link Beams in a Double-column Tall Pier Bent Subjected to Earthquake Excitations. *Earthq. Eng. Struct. Dyn.* **2023**, *52*, 641–659. [CrossRef]
26. Baei, M.; Terzic, V. Optimal Design of Dampers in Seismic Applications Utilizing the MOPSO Algorithm. *Front. Built Environ.* **2022**, *8*, 1040129. [CrossRef]
27. Sun, C.; Wang, K.; Liu, Q.; Wang, P.; Pan, F. Machine-Learning-Based Comprehensive Properties Prediction and Mixture Design Optimization of Ultra-High-Performance Concrete. *Sustainability* **2023**, *15*, 15338. [CrossRef]
28. Guo, W.; He, Y.; Zhu, Y.; Hu, Y. Performance Parameter Design of Seismic Isolation Bearings for High-Speed Railway Simply-Supported Bridges Using Neural Network. *Soil Dyn. Earthq. Eng.* **2024**, *179*, 108535. [CrossRef]

29. Fang, C.; Ping, Y.; Gao, Y.; Zheng, Y.; Chen, Y. Machine Learning-Aided Multi-Objective Optimization of Structures with Hybrid Braces—Framework and Case Study. *Eng. Struct.* **2022**, *269*, 114808. [CrossRef]
30. Jiang, M.; Rui, X.; Yang, F.; Zhu, W.; Zhang, Y. Multi-Objective Optimization Design for a Magnetorheological Damper. *J. Intell. Mater. Syst. Struct.* **2022**, *33*, 33–45. [CrossRef]
31. Yang, G.; Zhang, T.; Mao, J.; Tian, L.; Du, Y. A NSGA-II-Based Approach for Optimizing Structural Component Pre-Reinforcement to Enhance Cable-Stayed Bridge Resilience. *Adv. Struct. Eng.* **2024**, *28*, 1075–1092. [CrossRef]
32. Tian, Z.; Zhang, Z.; Ning, C.; Peng, T.; Guo, Y.; Cao, Z. Multi-Objective Optimization of Cable Force of Arch Bridge Constructed by Cable-Stayed Cantilever Cast-in-Situ Method Based on Improved NSGA-II. *Structures* **2024**, *59*, 105782. [CrossRef]
33. Almutairi, A.; Lu, J.; Wang, N.; Elgamal, A. *Analysis of Multi-Span Bridges Using OpenSees*; National Academies: Washington, DC, USA, 2016.
34. Rastgoo Moghadam, S.; Konstantinidis, D. Experimental and Analytical Studies on the Horizontal Behavior of Elastomeric Bearings under Support Rotation. *J. Struct. Eng.* **2021**, *147*, 04021024. [CrossRef]
35. Kalfas, K.N.; Ghorbani Amirabad, N.; Forcellini, D. The Role of Shear Modulus on the Mechanical Behavior of Elastomeric Bearings When Subjected to Combined Axial and Shear Loads. *Eng. Struct.* **2021**, *248*, 113248. [CrossRef]
36. Joyner, M.D.; Gardner, C.; Puentes, B.; Sasani, M. Resilience-Based Seismic Design of Buildings through Multiobjective Optimization. *Eng. Struct.* **2021**, *246*, 113024. [CrossRef]
37. Deutsch, J.L.; Deutsch, C.V. Latin Hypercube Sampling with Multidimensional Uniformity. *J. Stat. Plan. Inference* **2012**, *142*, 763–772. [CrossRef]
38. Akiba, T.; Sano, S.; Yanase, T.; Ohta, T.; Koyama, M. Optuna: A Next-Generation Hyperparameter Optimization Framework 2019. In Proceedings of the 25th ACM SIGKDD International Conference, Anchorage, AK, USA, 4–8 August 2019.
39. Verma, S.; Pant, M.; Snasel, V. A Comprehensive Review on NSGA-II for Multi-Objective Combinatorial Optimization Problems. *IEEE Access* **2021**, *9*, 57757–57791. [CrossRef]
40. Jia, H.; Liu, Z.; Xu, L.; Bai, H.; Bi, K.; Zhang, C.; Zheng, S. Dynamic Response Analyses of Long-Span Cable-Stayed Bridges Subjected to Pulse-Type Ground Motions. *Soil Dyn. Earthq. Eng.* **2023**, *164*, 107591. [CrossRef]
41. *JTG D20—2017*; Specifications for Seismic Design of Highway Bridges. Standards Press of China: Bejing, China, 2020.
42. Kamrava, A. Comparing Results of MATLAB and Seismosignal in Plotting Earthquake Graphs. *Curr. World Environ.* **2015**, *10*, 11–14. [CrossRef]
43. Ras, A.; Boumechra, N. Seismic Energy Dissipation Study of Linear Fluid Viscous Dampers in Steel Structure Design. *Alex. Eng. J.* **2016**, *55*, 2821–2832. [CrossRef]
44. Liu, Q.; Zhu, S.; Yu, W.; Wu, X.; Song, F.; Ren, X. Ground Motion Frequency Insensitivity of Bearing-Supported Pedestrian Bridge with Viscous Dampers. *KSCE J. Civ. Eng.* **2021**, *25*, 2662–2673. [CrossRef]
45. Panda, S.K.; Jana, P.K. Efficient Task Scheduling Algorithms for Heterogeneous Multi-Cloud Environment. *J. Supercomput.* **2015**, *71*, 1505–1533. [CrossRef]
46. Marler, R.T.; Arora, J.S. The Weighted Sum Method for Multi-Objective Optimization: New Insights. *Struct. Multidisc. Optim.* **2010**, *41*, 853–862. [CrossRef]
47. Yu, C.; Xiang, H.; Li, Y.; Pan, M. Optimization of Longitudinal Viscous Dampers for a Freight Railway Cable-Stayed Bridge under Braking Forces. *Smart Struct. Syst.* **2018**, *21*, 669–675.
48. Ozdagli, A.; Volgyesi, P.; Koutsoukos, X. Surrogate Modeling Using Physics-Guided Learning. In Proceedings of the Cyber-Physical Systems and Internet of Things Week 2023, San Antonio, TX, USA, 9 May 2023; ACM: New York, NY, USA, 2023; pp. 130–135.

www.ingramcontent.com/pod-product-compliance
Lightning Source LLC
Chambersburg PA
CBHW062046220926
51850CB00010B/2682
* 9 7 8 3 7 2 5 8 4 8 3 9 3 *

MDPI AG
Grosspeteranlage 5
4052 Basel
Switzerland
Tel.: +41 61 683 77 34

Buildings Editorial Office
E-mail: buildings@mdpi.com
www.mdpi.com/journal/buildings